Bonding by
Self-Propagating Reaction

David J. Fisher

Published by **Materials Research Forum LLC**
Millersville, PA 17551, USA

Published as part of the book series
Materials Research Foundations
Volume 45 (2019)
ISSN 2471-8890 (Print)
ISSN 2471-8904 (Online)

Print ISBN 978-1-64490-008-6
ePDF ISBN 978-1-64490-009-3

Distributed worldwide by

Materials Research Forum LLC
105 Springdale Lane
Millersville, PA 17551
USA
http://www.mrforum.com

Printed in the United States of America
10 9 8 7 6 5 4 3 2 1

Table of Contents

Materials Research Forum LLC
doi: http://dx.doi.org/10.21741/9781644900093

Bonding by Self-Propagating Reaction

Just as the transient liquid phase bonding technique[1] can be argued to have its roots in ice-sculpting, where a sprinkling of salt could join blocks of ice, so too can the self-propagating reaction bonding method be said to have its roots in the thermite process for repairing rail-track. As in the case of transient liquid phase bonding, reactive multilayer joining again involves sandwiching an interlayer between the materials to be joined. The interlayer here, rather than forming a lower melting-point alloy with the adjoining materials, is instead a bimetallic multilayer which can generate an amount of localised heat that is sufficient to melt the adjacent materials: sometimes those materials themselves or more probably a solder-like filler.

One specific patented example[2], involving the Ti|B system, envisages a 5mm x 5mm silicon piece placed atop a 10mm x 10mm silicon piece (figure 1), both having flat surfaces polished to a smoothness of $1\mu m$. Onto each surface is then evaporated, or sputter-deposited, 500Å of titanium, chromium or zirconium. Onto those surfaces are vacuum-coated $1.75\mu m$ of aluminium. Alternating layers of 0.5 to $50\mu m$ of titanium and boron are subsequently vacuum-coated onto the aluminium layers. A second $1.75\mu m$ aluminium layer is vacuum-coated onto each of the titanium/boron multilayers. The two coated pieces are then squeezed together in a suitable jig.

Reaction of the component elements of the interlayered coatings is initiated by, for example, a high-voltage spark and this produces molten aluminium plus reaction products. The temperature which is attained due to the heat output of the reaction is sufficient to melt the aluminium but does not appreciably affect the remainder of the assembly. That is, the inherently minute mass of the reactants severely limits its thermal output. The aforementioned patent explains this in some detail: one mole of titanium combines with two moles of boron to give one mole of TiB_2; the heat of reaction being between 71 and 85kcal/mol, or some 1100cal/g. One mole of titanium is equivalent to $10.55cm^3$ and two moles of boron are equivalent to $9.12cm^3$, giving $15.44cm^3$ of TiB_2. The total volume of 53.6% titanium and 46.4% boron thus decreases from 19.67 to $15.44cm^3$. This tends to be compensated by the infiltration of boride cracks by molten aluminium.

Materials Research Forum LLC
doi: http://dx.doi.org/10.21741/9781644900093

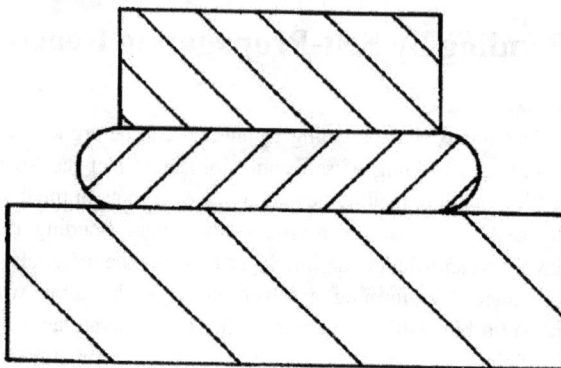

Figure 1. Generic joining arrangement using a reactive multilayer

A 1μm x 1cm x 1cm multilayer with that ratio of unreacted elements consists of 0.53μm of titanium and 0.46μm of boron; implying the presence of 0.00024g of titanium and 0.00011g of boron, giving a total weight of 0.0035g. Thus, given the above figure of 1100cal/g, such a deposit can provide some 0.386cal. The temperature-rise can then be calculated by assuming that only aluminium and TiB_2 are present. The heat capacity of solid aluminium is $(4.94 + 0.00296T)$cal/mol and that of the boride is 0.45cal/g, so that the heat capacity of the 1μm x 1cm x 1cm multilayer is 0.0001575cal/deg. The heat capacity of solid aluminium is about 6.757cal/mol, assuming 26.98g/mol, and the 3.5μm x 1cm x 1cm layer weighs 0.000944g, assuming a density of 2.698g/cm³. Its heat capacity is therefore 0.000236cal/deg, and the heat capacity of the whole system is 0.0003936cal/deg. In order to heat it from 298K to 931.7K, the melting-point of aluminium, thus requires 0.249cal. The heat of fusion of aluminium is 2570cal/mol (95.2cal/g): equating to 0.09cal for a 4μm x 1cm x 1cm aluminium layer. The overall energy required in order to heat the aluminium and TiB_2, and to melt the aluminium, is 0.339cal. Upon comparing this value with the 0.386cal calculated above for the output of the reactive foil layer, it is seen that the aluminium can be melted but also that there is little extra heat available to affect the joined components. The difference of 0.047cal in fact increases the temperature further but, given that the total heat capacity of the 3.5μm x 1cm x 1cm aluminium layer and of the TiB_2 is 0.0004cal/deg, the additional temperature increase is only about 100C.

Materials Research Forum LLC
doi: http://dx.doi.org/10.21741/9781644900093

The above is of the nature of a generic computation intended to illustrate the various phenomena at issue but the principle is not limited to the use of titanium, boron and aluminium. The multilayer film can instead comprise zirconium, hafnium, niobium or tantalum, alternating with carbon or silicon, and the molten material can be tin, copper, silver, germanium or silicon. The materials which can be joined in this way include, in addition to silicon, a wide range of metals, intermetallics, metalloids and ceramics. The low melting-point filler metal or alloy is placed between the reactive-layer heat source and the items to be joined in order to encourage bonding between those items and the reactive multilayer. With access to modern coating technology, reactive multilayer bonding is thus relatively simple and rapid because it does not require the use of furnaces, or even pre-heating, and can be performed in air.

The present work is restricted to the presentation of reactive-foil bonding. The method is partly rooted however in the use of powders; as in the classic, 'rust plus aluminium filings' thermite technique. An early patent[3] describes a process in which "*an entire reaction mass contained in a melting vessel is caused to react at one point, the reaction continuing to propagate endothermically from this base*". This led on to the study[4] of reaction propagation in layers of compacted powder. The concept of using metallic foils perhaps owes more however to even earlier studies[5] of rapid phase changes in metalloid layers such as, for example, antimony. In a much later study[6] two red-hot needles were dropped, from different heights, onto polished specimens bearing a thin antimony layer. This explosive electrolytic deposit had an heterogeneous gel-like structure with one phase oriented parallel to the path of the deposition current. The speed of the explosive wave was then determined by measuring the distance, from the two craters which had been made a small known fraction of a second apart, to the line of meeting of the two sets of waves. The rate of phase-change propagation was found to range from 10 to 40cm/s, and was a function of the temperature and the thickness of the antimony deposit. It is now known that so-called explosive crystallization of amorphous films requires films of between 2 and 100μm in thickness, and this requirement is attributed to the relatively small heat of crystallization.

The traditional thermite formulation is after all nothing more than a deflagratory explosive in which rust is the oxidant and aluminium powder is the fuel. It is no coincidence that aluminium is a component of many of the materials which support self-propagating reaction. One might almost imagine that some sort of 'anthropic principle' is at work when one contemplates the coincidences associated with aluminium. It is hard to imagine how the modern world could have advanced without that metal, and it is thus a happy accident that it just so happens to be the most common metal in the Earth's crust. On the other hand, it could not be exploited until relatively recent times because of its

Materials Research Forum LLC
doi: http://dx.doi.org/10.21741/9781644900093

extreme affinity for oxygen and the consequent enormous difficulty of extracting it from its oxide-ore, bauxite. Its extreme affinity for oxygen is often forgotten until a foundry is destroyed by pouring molten aluminium into a damp mould, or the aluminium superstructure of a warship is ignited by an enemy missile.

The fact that aluminium articles do not therefore routinely fizz away like sodium is due to another happy coincidence: the metal possesses a perfect Pilling-Bedworth Ratio. That is, its oxide neither fails to prevent further oxidation by leaving gaps in its coverage nor does it crack and flake off because the oxide is too voluminous.

The self-propagating exothermic reactions which are to be discussed here can be ignited at room temperature, and bonding can be completed at room temperature, in air or other gas, in less than one second. The use of reactive foils obviates any need for furnaces, and markedly reduces the total heat budget that is required. Temperature-sensitive components can be joined without suffering thermal damage. Mismatches in thermal contraction during cooling, as may occur when joining metals to ceramics, are avoidable because of the very small increases in temperature which are involved. The assembly to be bonded may also be compressed uniaxially using pressures of between 0.1 and 10MPa.

The concept of self-propagating reaction bonding can be considered to be a special case of a technique known as combustion synthesis. Since the 1970s, this has become an increasingly important technique; especially for preparing ceramic composites, being very quick and low-energy when compared with isostatic pressing and sintering. This is because it exploits the ability of highly exothermic chemical reactions to become self-sustaining. The method can be divided into two branches. In the so-called self-propagation mode, the assembly is heated locally to the ignition temperature and the heat generated by the chemical reaction then raises adjacent layers to the ignition temperature. This instigates a self-sustaining combustion wave. In the so-called simultaneous mode, the sample is heated uniformly to the ignition temperature. This leads to rapid exothermic conversion of the reactants.

Like the original thermite composition, one class of these reactive materials is based upon a metal and an oxide. As well as their use for bonding materials, as considered here, they may also be used as additives to conventional propellants, explosives and pyrotechnics. Their use, especially for bonding purposes, requires the reliable predictability of their behaviour as a function of heating-rate, pressure and environment. Much research is carried out on quantifying the ignition temperatures and kinetics by using simple configurations and low heating-rates. Large discrepancies continue to exist between experimental data and the predictions of ignition models. Thermo-analytical

Materials Research Forum LLC
doi: http://dx.doi.org/10.21741/9781644900093

measurements have been used to identify and describe quantitatively the reactions which lead to ignition.

Thermite-Type Bonding

The class of thermite analogues, which is ultimately the least favoured for most bonding operations, will be considered first, the main point of interest always being the range of materials which can in fact be bonded.

The common feature of most of these pairs is a large heat of formation and high reaction temperature, with the underlying driving force being the usual tendency of a system to minimize its free energy. Again as usual, other factors militate against this tendency. These include the time required for mass and heat diffusion. The layers can react explosively or via self-sustained propagation, in the manner of a fuse.

The former situation occurs if the temperature of the reactants is increased rapidly to the so-called ignition temperature, whereupon there is an homogeneous reaction which involves simultaneous reaction of all of the constituents.

In the latter situation, again like a fuse, some form of energy is applied to one point and triggers a self-propagating reaction. That is, the locally applied energy causes the reaction of a minute portion of the reactants. The resultant burst of heat diffuses into the rest of the film, causing further reaction and heating and sustaining a wave of reaction. An important point is that the required heat energy may be instead lost to the adjacent material, thus impeding continued reaction. Rapid heat transfer through the reactive film is thus essential to the maintenance of propagation. As noted above, the reactive system can be metal|metal, metal|semiconductor, metal|metalloid or metal|oxide. Metal|organic multilayers are also feasible, but will not be considered further in the present work.

The reaction-product can comprise the original reactants, as in what might be called a 'metallurgical' reaction (e.g. Ni+Al→NiAl), or involve a true chemical (reduction/oxidation) reaction, as in the classic thermite formulation. Although the propagation rates can be fairly high, the reactions are characterised by deflagration rather than detonation. That is, the reaction wave moves at subsonic speeds. Detonation would involve a reaction wave which exceeded the local speed of sound. Mass transport is a major rate-limiting process during self-propagating reactions.

Vacuum deposition is an important technique in the present field, because it creates a well-defined heterostructure having a controlled interfacial area and minimal void content. Joining operations require multilayer heat sources which can comprise perhaps thousands of individual layers, and the total interfacial area is indirectly governed by the choice of layer thickness and the number of layers. Vacuum-based deposition methods here provide a perfect solution for thin-film growth and the use of nanometre dimensions.

$Al|Co_3O_4$

Ferromagnetic $Co|Al_2O_3$ nanocomposite thin films have been prepared by the thermite reaction of aluminium and Co_3O_4 in a layer geometry which was obtained by depositing aluminium layers onto Co_3O_4 films at room temperature and annealing at between 50 and 700C in 50C steps[7]. At above about 450C, the Co_3O_4 partially transformed into CoO. The simultaneous solid-state reaction of aluminium with Co_3O_4 and of aluminium with CoO began above an initiation temperature of some 500C, regardless of the bilayer thickness. Following annealing at 700C, about 60% of the cobalt was reduced by the aluminium while the remainder of the cobalt was present as intermediate $CoAl_2O_4$ shells which demarcated the cobalt nanoparticles from the Al_2O_3 matrix. At above 700C, the reaction was complete and the final products consisted of non-interacting cobalt nanoparticles, with an average size of about 40nm, enveloped by $CoAl_2O_4$ shells and embedded in a dielectric Al_2O_3 matrix.

$Al|CuO$

Sputter-deposited multilayers of this type currently represent the state of the art of thermite-type energetic nanomaterials. The theoretical energy density of such combinations is appreciably higher than that of most conventional secondary explosives, and they are less sensitive to accidental initiation. The reaction temperature, reaction velocity, reaction products and sensitivity can be manipulated by controlling the layering, reactant spacing and ratio of the multilayers. This is best done by direct-current sputter deposition[8].

In one study, a reactive multilayer foil was studied which involved the classic reduction-oxidation thermite reaction between a copper oxide and aluminium. Rather than forming an intermetallic compound, as in the case of the class of materials to be considered later, the aluminium and copper oxide were oxidized and reduced, respectively, to give aluminium oxide and copper[9]. Thermite foils containing CuO_x and aluminium were sputter-deposited to form a multilayer. The CuO_x crystal structure was the same as that of paramelaconite, Cu_4O_3; other methods having confirmed that the oxide was richer in copper than was the CuO sputter target. Oxygen was found to be homogeneously dispersed throughout the CuO_x layer, but the concentration of oxygen in any given aluminium layer was minimal because the only oxygen contribution arose from the surface. A narrow region in the interface between the two layers was made up of amorphous and nanocrystalline Al_2O_3. Differential thermal analysis showed that the

CuO$_x$|Al foils reacted in the expected highly exothermic manner and that two major exotherms appeared during heating. The total heat released, -3.9kJ/g, was similar to the heat-of-reaction which was predicted for the reaction of CuO and aluminium. This amount of heat was sufficient to cause the reaction to self-propagate through the foil at a velocity of 1m/s. The reaction temperature was at least as high as the evaporation temperature (2846K) of copper. On the basis of the reaction velocities of CuO$_x$|Al and Ni|Al foils, it was expected that the kinetics of the CuO$_x$|Al reaction would be much slower than the kinetics of Ni|Al formation; Ni|Al being a leading example of the rival class of self-propagating reaction bonding materials.

The reaction of such multilayer thin films was also studied by varying the substrate material and thickness, with the in-plane speed of propagation of the reaction being determined by using a time-of-flight technique. The reaction was completely quenched by a silicon substrate with an intervening silica layer of less than 200nm[10]. The speed of reaction was constant at 40m/s for silica layers which were thicker than 1μm. In general, the reaction speed ranged from 2 to 61m/s. In another study, all of the energy was released below the melting point of aluminium[11]. In microstructured samples, at least two thirds of the energy was released between 1036 and 1356K. It was confirmed that microstructured samples decomposed in two-step reactions: there was an initial exothermic reaction of 0.7kJ/g at 790K, and a second exothermic reaction (1.3kJ/g) between 1036 and 1356K. On the other hand, there was a single exothermic reaction (1.2kJ/g) at about 740K; well below the melting point (933K) of bulk aluminium. The stresses in the foils depended upon the deposition conditions (table 1), and had an effect upon the density and porosity of the deposited layers.

Table 1. Stresses in deposited Al|CuO foils

Total Thickness (μm)	CuO Thickness (nm)	Al Thickness (nm)	Stress (MPa)
1.1	50	25	11
1.15	50	50	18
1.0	100	100	42
1.1	100	100	48
2.1	100	100	24
2.1	1000	1000	29
3.0	1000	1000	17

Density functional theory computations[12] have suggested that prior intermixing occurs via aluminium penetration into the CuO surfaces and the penetration of copper and oxygen atoms into the aluminium following the dissociative chemisorption of CuO.

The heat released during the thermite reaction between aluminium and CuO films has been found to be 1537J/g. Reaction-temperature plots reach a peak in about 0.5µs, and then fall to a relatively stable temperature[13]. Certain layer-structure Al/CuO multilayers had a reaction heat of 2760J/g, and the reaction temperature of over 3500K could persist for 2.4ms[14,15]. Copper and Cu|Al|CuO multilayer films were produced by using standard and radio-frequency magnetron sputtering, respectively[16]. The as-deposited Al|CuO multilayer films had a distinct layer structure. Differential scanning calorimetry here showed that the amount of heat which was released was 2024J/g. A comparison with the copper film revealed that large numbers of product particles were violently ejected to a distance of some 6mm from Cu|Al|CuO multilayers. The reaction temperature was 6000 to 7000K and 8000 to 9000K for copper film and Cu|Al|CuO multilayer films, respectively.

A two-dimensional diffusion-reaction model was used to predict the ignition threshold and reaction dynamics of Al|CuO multilayer thin films, while taking account of the fact that CuO first decomposed into Cu_2O while the released oxygen diffused across the Cu_2O and Al_2O_3 layers before reacting with pure aluminium to form Al_2O_3. The model was experimentally confirmed by using ignition and reaction velocity data on Al|CuO multilayers[17]. In this system, reaction is thus dominated mainly by the outward migration of oxygen atoms from the CuO matrix and towards the aluminium layers. The interfacial nanolayer between the two reactive layers therefore plays a major role in determining the overall properties. The atomic-layer deposition of a thin ZnO layer onto CuO before aluminium sputter-deposition led to a marked increase in the efficiency of the overall reaction[18]. The CuO|ZnO|Al foils generated 98% of the theoretical enthalpy during a single reaction at 900C. Normal ZnO-free CuO|Al foils produced only 78% of the theoretical enthalpy and required two distinct reaction steps at 550 and 850C. The successive formation of thin $ZnAl_2O_4$ and ZnO interfacial layers acted as effective barrier layers to oxygen diffusion at low temperatures. Ignition of Al|CuO multilayers was studied experimentally, showing that the heating surface area has to be properly chosen in order to optimize the nanothermite ignition: a heating surface area of $0.25mm^2$ was sufficient to ignite a multilayered thermite film which was 1.6mm wide and some cm long. A new analytical model, based upon atomic diffusion across the layers and upon thermal exchange, was proposed which assumed – as above - that CuO first decomposes to give Cu_2O and that the oxygen then diffuses across the Cu_2O and Al_2O_3 layers before reaching the aluminium layer and reacting to form Al_2O_3. An increase in the heating-

Materials Research Forum LLC
doi: http://dx.doi.org/10.21741/9781644900093

surface area led to an increase in the ignition response time and ignition power threshold. For a given heating-surface area, the ignition time decreases rapidly and asymptotically as the electrical power density increases. When the aluminium thickness is half of that of the CuO thickness, the minimum ignition response time can be varied from 59μs to 418ms by adjusting the heating-surface area[19]. The minimum ignition response time increases with increasing bilayer thickness.

The ignition characteristics of Al|CuO nano-energetic multilayer films, integrated with a semiconductor bridge, were investigated[20]. Distinct films were sputter-deposited in layer form, and exhibited a heat-of-reaction of 2181J/g. The films did not affect the electrical properties of the semiconductor bridge.

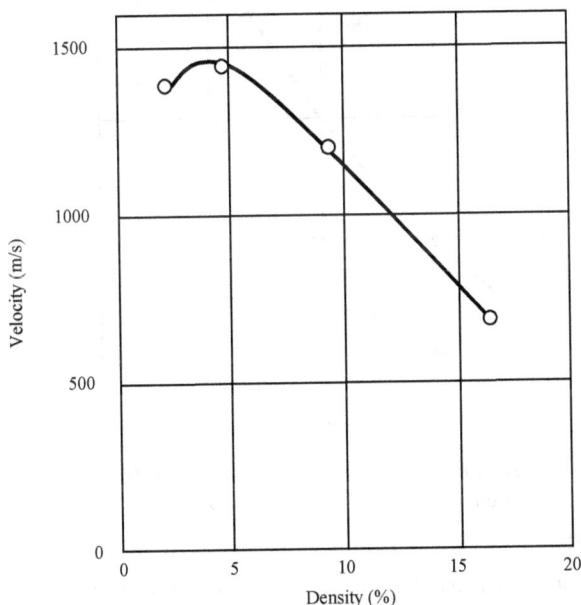

*Figure 2. Combustion velocity as a function of
the theoretical density of Al|CuO composite*

Heating samples at a low controlled rate within a differential thermal analyzer shows that the reduction of CuO_x and the oxidation of aluminium indeed involves two distinct exotherms[21]. Samples were heated to various temperatures within these exotherms,

Materials Research Forum LLC
doi: http://dx.doi.org/10.21741/9781644900093

quenched and examined thus revealing that - in the first reaction - CuO_x is reduced to a mixture of CuO and Cu_2O while an interfacial layer of Al_2O_3 grows and coalesces. The final products of the second exotherm were copper, Al_2O_3 and Cu_2O. The first exotherm was attributed to two-dimensional interface-limited growth of the Al_2O_3 layer. The second exotherm was attributed to diffusion-limited one-dimensional growth of Al_2O_3 and to the interface-controlled growth of copper involving the reduction of Cu_2O.

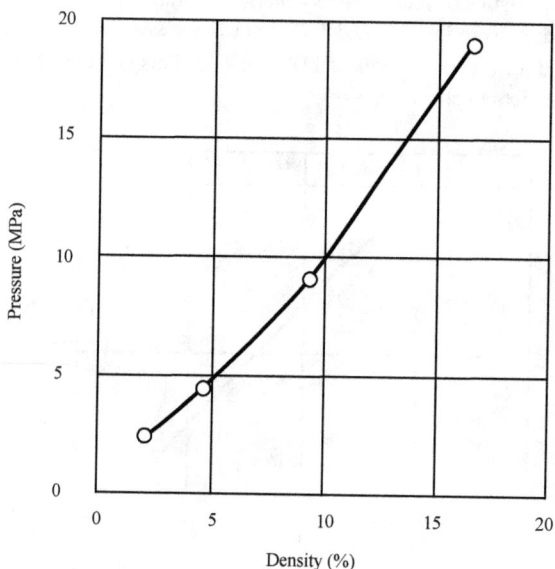

Figure 3. Peak pressure as a function of the theoretical density of Al|CuO composite

Higher combustion front speeds had been observed in composites with ordered porous Fe_2O_3, as an oxidizer, and aluminium fuel nanoparticles, as compared with composites which contained a porous oxidizer with no ordering of the pores and nanoparticles. Rather than layering the components of the present thermite, it was therefore made by synthesizing CuO nanorods via the surfactant-templating method and then mixing or self-assembling with aluminium nanoparticles[22]. Such nanoscale mixing resulted in a large interfacial contact area, leading to a rapid reaction which could generate shock waves having Mach numbers of up to 3 (figure 2). The composite exhibited a combustion front

Materials Research Forum LLC
doi: http://dx.doi.org/10.21741/9781644900093

speed of 1500m/s, which increased to 2200m/s for the self-assembled composite. A similar technique was to synthesize CuO nanowires by thermally annealing copper film which had been deposited onto silicon[23]. Nano-aluminium was then integrated with the nanowires so as to produce the Al||CuO thermite. These high combustion-front speeds approached the lower limits of the detonation velocities observed in conventional explosives but, because of the low density and multiphase nature of the reactive materials, the Chapman-Jouguet pressure (figure 3) could be much lower than that of conventional solid explosives.

A room-temperature reaction-bonding method for aluminium has been proposed which involves combining the Al||CuO powder thermite reaction with the self-propagation reaction of Ti-B powder, with no external heat source. The joint microstructure comprises Al_2O_3, TiB_2 and (Al,Cu) intermetallics, whose distribution governs the joint properties. The multiple reactions produce a compact joint layer with very low porosity, and this is attributed to the flow of molten aluminium metal into the joint layer during joint formation[24]. Bonding between the various components of the joint is due to lattice defects at their interfaces. Shear tests show that fracture occurs at the metal||joint interface, where Al_2O_3 particles segregate.

Due to the tendency of these thermite-type layers to produce gaseous products, such so-called redox foils have to be 'diluted' in order to avoid metal-vapor formation. This however can lead to foils which react too slowly to be able to propagate fully during bonding. In order to improve the reactivity the aluminium, oxide and a copper diluent may be milled prior to mechanical consolidation into a dense redox foil[25]. Milling decreases the distance between the aluminium fuel and the oxidant, thus increasing oxygen transfer. In some cases, the aluminium and oxide thermite components are milled without the copper; which is then added before consolidation. In other cases, all of the constituents are milled together. Both choices produce foils wherein the reaction propagates faster and releases heat earlier than in foils made from non-milled powders. On the other hand, only the method in which all of the components are milled together avoids the production of copper vapor during reaction; the absence of such vapor being attributed to the more uniform distribution of copper. In order to study the precise mechanism of vapor production in such diluted thermites, thin magnetron-sputtered multilayer Al||Cu||Cu$_2$O||Cu foils were prepared in which the copper layer thickness was varied from 0 to 100nm in steps of 25nm[26]. The excess copper acted as a diffusion barrier which limited the transport of oxygen from the oxide to the aluminium. By adding excess copper, the temperature of the self-propagating thermite reaction was also caused to fall below the boiling-point of copper; thus avoiding metal-vapor production. It was found that copper-vapor generation could be avoided by increasing the copper interlayer thickness to above 50nm. Unfortunately the final product remained porous and it was concluded that, although metal-vapor generation could be avoided by dilution, the same could not be said of oxygen generation.

Reactions between metals such as aluminium, and transition-metal oxides, are of general interest due to their rapid reaction: of the order of 10μs. The extent of reaction in aluminium-based thermites has been investigated by examining the products formed by the nanothermites, Al‖CuO, Al‖Bi$_2$O$_3$ and Al‖WO$_3$, which have differing reaction characteristics. Microgram-sized samples were coated onto fine platinum wire and resistively heated at 10^{5}K/s in order to cause ignition. The product phases were preserved within 500μs in order to preserve their chemistry and morphology. It was found that oxygen was localized mainly in regions which contained aluminium; as expected for a redox reaction. The Al‖CuO system exhibited simultaneous gaseous oxygen release and ignition, with a much lower oxygen content in the products as compared with Al‖Bi$_2$O$_3$ or Al‖WO$_3$. The latter were suggested to undergo condensed-phase reaction. Small particles tended to have a higher oxygen content than did large ones. A notable result was that thermites which produce large amounts of gas produce smaller particles and enjoy greater degrees of completion.

Table 2. Velocity and duration of combustion fronts in Al‖CuO at various temperatures

Temperature (K)	Front	Velocity (m/s)	Duration (ps)
830	upper	335	0.43
1000	upper	127	1.70
2000	upper	201	1.09
3000	upper	430	0.85
830	lower	196	0.85
1000	lower	194	1.23
2000	lower	172	1.38
3000	lower	399	0.83

It has been demonstrated that inkjet-printing can be used for the deposition of energetic materials[27]. Largely inert colloidal suspensions of nano-aluminium and nano-copper oxide, in dimethylformamide with polyvinylpyrrolidone, were sequentially deposited onto a substrate by means of piezo-electric ink-jet printing. The arrangement was such

Bonding by Self-Propagating Reaction
Materials Research Foundations **45** (2019)

Materials Research Forum LLC
doi: http://dx.doi.org/10.21741/9781644900093

that the aluminium and copper oxide droplets were interspersed and overlapped; permitting *in situ* mixing of the components. Alternating deposition was repeated in order to create samples having multiple layers of energetic material. Samples which were printed with 3, 5 or 7 layers of material were then spark-ignited. It was noted that the maximum reaction temperature of samples which were printed by using a dual-nozzle method was some 200K lower than that for samples which were printed by using a single nozzle, although the degrees of mixing were comparable in both cases.

In another technique[28], electrophoretic deposition was used to produce 10 to 200μm-thick films of nano-aluminium|copper-oxide thermite compositions having the theoretical maximum density of 29%. The optimum reaction propagation velocity was found to correspond to a fuel-rich equivalence ratio of 1.7. This value did not correspond to the predicted maximum of gas production or temperature. A 25% decrease in propagation velocity occurred above the equivalence ratio of 2.0, where Al_2O_3 was predicted to change from liquid to solid. This was expected to slow the kinetics by decreasing the mobility of the condensed-phase reactants. The propagation velocity depended upon the film thickness: the velocity exhibited a two-plateau behavior in which one plateau – with a 4m/s velocity - was found between 13 and 50μm and the other – with a 36m/s velocity - was found above 120μm. A linear transitional regime existed in the 50 to 120μm range. There was an increase in the forward-transported particles as the film thickness increased, together with more turbulent behavior. It was proposed that the two-plateau behavior reflected a change in the energy transport mechanism. The velocity of the particles which were ejected through a thin slit, mounted above a thermite strip, was measured, and was found to be two or three times faster than the propagation velocity. A non-dimensional number was used to relate the rate of gas pressurization to the rate of gas-escape via Fickian diffusion. At small values of the number, gases escaped rapidly and did not accumulate within the thermite film so that the resultant energy transport was relatively slow. At large values of the number, the gases were completely trapped and this led to increased energy-transport via oscillating pressure build-up and unloading of the material.

Simulations which were based upon *ab initio* molecular dynamics methods have been used[29] to study the reaction of Al|CuO thermite at 830, 1000, 2000 and 3000K. This revealed that the redox reactions which produced copper metal and aluminium oxide began at both the upper and lower interfaces. The atomic configuration changed more rapidly with increasing temperature. At above 2000K, the copper began to penetrate into the aluminium layer and the copper was eventually diffused with aluminium atoms at 3000K. There were appreciable differences between the propagation rates and duration time at the upper and lower oxidation fronts. At 1000K, the rates of both oxidation fronts

Materials Research Forum LLC
doi: http://dx.doi.org/10.21741/9781644900093

were lower than for others (table 2). At 3000K, the rate attained a maximum value of 430m/s in the case of the upper oxidation front. A fast thermite reaction occurred at higher temperatures. At 830 and 1000K, the reactions were incomplete. The reaction finished at 2000K. At 3000K, the copper migrated into the aluminium oxide.

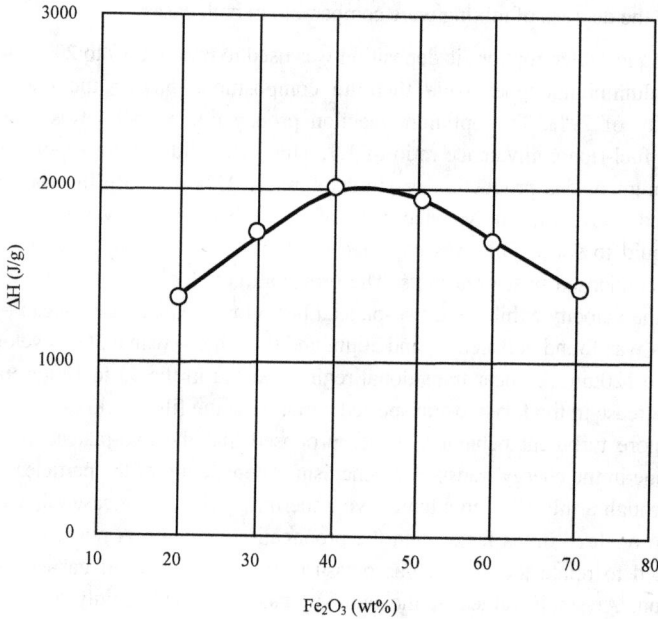

*Figure 4. Total thermal energy released by Al|Fe$_2$O$_3$
layers as a function of the oxide content*

Al|Fe$_2$O$_3$

Among the thermite-type reactive multilayers, it is of course not surprising to find the 'primordial' composition involving aluminium and an iron oxide. The effect of aluminium and iron oxide in the form of micro- and nano-scale particles as a thermal energy source for melting solder microparticles (SAC305) for the bonding of copper substrates was investigated[30]. The optimum mixture was found to consist of aluminium microparticles, aluminium nanoparticles and Fe$_2$O$_3$ nanoparticles in the weight ratio of

Bonding by Self-Propagating Reaction   Materials Research Forum LLC
Materials Research Foundations **45** (2019)   doi: http://dx.doi.org/10.21741/9781644900093

30:30:40. This generated a maximum total exothermic energy of about 2.0kJ/g (figure 4). The inclusion of aluminium nanoparticles was essential in order to ensure the stable ignition and initiation of aluminium microparticles and thus maintain a relatively long duration and rate (figure 5) of combustion.

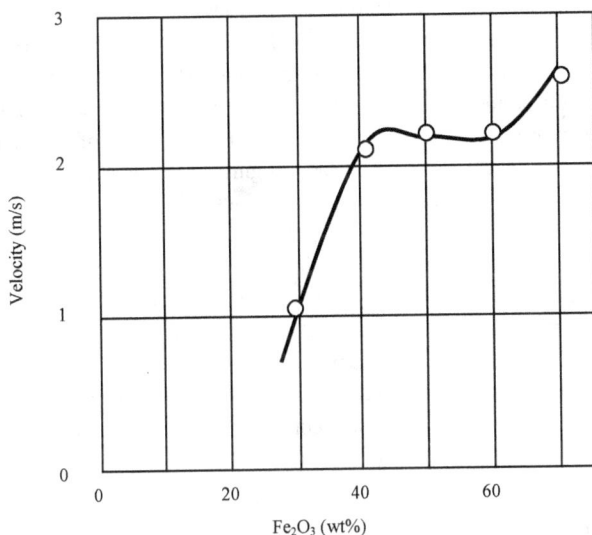

*Figure 5. Reaction rate of Al\Fe₂O₃ composite
pellets as a function of the oxide content*

Use of the highly reactive aluminium and oxide nanoparticles could improve the aluminothermic reaction, although the addition of aluminium microparticles to those nanoparticles was required in order to maintain a high thermal energy for longer times. An energetic layer comprising aluminium microparticles, aluminium microparticles, aluminium nanoparticles and Fe_2O_3 nanoparticles as a composite was used as a heat source between SAC305 microparticle solder layers. The solder|layer|solder multilayer pellets were then ignited between copper substrates. Successful interfacial bonding was achieved and the maximum mechanical strength of the bonded copper substrates produced in this manner was about 40% higher when compared with the result of using solder alone. It was concluded that the reactive layers could serve both as an efficient thermal energy source and as mechanical reinforcement.

Materials Research Forum LLC
doi: http://dx.doi.org/10.21741/9781644900093

Al|MoO₃

The self-propagation reaction behavior of magnetron-sputtered reactive multilayer films with various modulation periods and widths, deposited onto glass substrates, was studied by means of differential scanning calorimetry[31]. This revealed that thermal reaction occurred in the solid|solid phase for modulation periods of 50 and 150nm and in the liquid|solid phase for a modulation period of 1500nm. The films underwent stable self-propagating reaction for modulation periods 50 to 150nm when excited by a laser pulse, but not for modulation periods of 300 to 1500nm. The average stable reaction velocity was 6m/s for a 150nm modulation period, and attained 10m/s for one of 50nm. The effect of the width upon the reaction velocity was insignificant. Three types of energetic semiconductor bridge were integrated by using various $Al|MoO_x$ energetic multilayer nanofilms[32]. The critical firing energy was found to be positively proportional to the modulation period of the nanofilm for a given initiation voltage, whereas the total firing energy and the input energy-use efficiency did not change significantly[33]. When samples having aluminium particle sizes of 44, 80 or 121nm, and bulk powder densities ranging from 5 to 10% of the theoretical maximum density, were tested the reaction velocity ranged from some 600 to 1000m/s. The velocity increased with decreasing particle size[34]. Pressure measurements showed that strong convection effects contributed greatly to reaction propagation.

Non-Thermite Type Bonding

This class of bonding material is remarkable because it is so unlike the thermite case, in that the later is easily understood as being a straightforward chemical reaction. It is often forgotten, when handling metals in bulk form, that they are nevertheless quite reactive towards each other. Their reactivity may not however become obvious unless they are put into a form where the surface-area to volume ratio is large. For example, lead, normally a very docile material will explode when ball-milled and then exposed to the atmosphere. This can again be seen as being a familiar chemical reaction. At the same time, the data in compilations such as Kubachewski's classic, *Metallurgical Thermochemistry*, clearly show that the reaction of two metals can easily release just as much energy as would burning either of them, lead-like, in air. In the present case, the reactivity can be exploited by using very thin films of the materials.

Nanostructured bimetallic reactive multilayers can be prepared via the ball-milling of elemental powders. Temperature increments during impactor collisions have been modelled by using Green's functions, and source images with respect to warped ellipsoid domain boundaries. The heat-source efficiency was deduced from laboratory data, taking account of thermal expansion and impactor inelasticity, and a thermal analysis was coupled to a dynamic mechanical model for particulate fracture[35]. This model yielded good agreement with the fractal dimension of the microstructures resulting from ball-milling.

A longer melting period can affect de-wetting and pore formation, while a shorter melting period tends to lead to inferior bond strength. The properties and quality of the joint depend strongly upon the duration of melting and the applied pressure and these parameters have to be optimised for each material. The use of vapor-deposition permits accurate control of the layer thickness and film composition; most multilayers being prepared using magnetron sputter-deposition or electron-beam evaporation. The multilayer ignition threshold, the reaction rate and the overall heat output can be tailored by means of careful thin-film design: planar multilayers of nm-scale periodicity exhibit self-sustained reactions having wave-front velocities of up to 100m/s. Some reactive multilayers produce product phases which are consistent with the relevant equilibrium phase diagram, while others produce metastable products.

Reactive foils are the only commercially available products at the moment, and their use is limited by their brittleness and limitation to planar geometries; reactive particles are more versatile but are not commercially available. Two-step electroless plating has been used for the synthesis of nickel-aluminium structures which act as micro-reactors and provide the energy required for joining. The energy input and particle size markedly affect the activation and reaction behavior of the structures[36]. In general, reactive foils range from 40 to 100μm in thickness and comprise many nanolayers that juxtapose materials having large heats of mixing, such as aluminium and nickel. A good rule-of-thumb guide to the required thickness is the expression,

$$\text{diffusion distance} = \sqrt{(Dt)}$$

where D is the diffusivity and t is an elapsed time. The diffusion-distance should obviously be commensurate with the film thickness and the rate-of-reaction. Because of the importance of solid-state diffusion, relevant data - where available - are included in the present work. By inserting a free-standing foil between two layers and two components, the heat generated by the reaction of the foil can be caused to melt the solder/braze and thus bond the components. So the use of a reactive foil eliminates the need for a furnace, and sharply reduces the degree of heating of the components to be bonded. Ceramics and metals can then be joined over large areas without introducing damaging thermal stresses. Pressure may be applied using a spring or a stationary plunger. The process is closely related to the so-called 'solid flame' concept[37], but the Ni|Ti system seems to be the only one which is deemed to be truly of that type[38].

Sputter deposition, electron-beam evaporation and, more recently, magnetron sputtering and atomic layer deposition have been used to prepare multilayer foils. Chemical vapor deposition is not likely to be suitable, because of the relatively high growth temperatures which are required. Vacuum deposition is generally preferred so as to minimize the inclusion of oxygen and other impurities. The multilayers are usually planar, with alternating layers of the two reactants; their combined thickness being termed the bilayer thickness. This thickness is usually constant throughout a given foil, but its make-up varies from system to system; being chosen so as to produce a required final composition. The ratio of the individual thicknesses which make up the bilayer thickness will therefore depend upon both the desired stoichiometry and the atomic densities of the two components. The correct composition for the multilayer can be closely approached if the above densities are known, but metalloid and oxide may be amorphous and thus require density measurements to be performed. It may also be desired to tailor the layer thicknesses so as to optimize other features, such as the final mechanical properties or the heat of formation generated during reaction. Sputter-deposition offers the possibility of graded and dual periodicity. Problems can arise when oxide layers are involved, as an adjacent metal layer may be prematurely oxidized during film growth.

Because ignition of the reactive multilayer is intended to be possible by using local impact or other energy-input, there is a distinct danger that such ignition may occur accidentally during multilayer preparation, when inspecting an assembly using various sorts of radiation or while carrying out final machining. Pre-mixing of the components over 0.1 to 10nm can occur due to diffusion during deposition, and have a marked effect upon performance because it can deplete the amount of stored chemical which is available for the generation of high temperatures, or create a diffusion barrier.

It is now known that a self-propagating reaction can be steady, pulsatile, 'spinning', chaotic or repetitive. When the titanium-nitrogen system was reacted at nitrogen pressures of 0.1 to 50kPa, planar surfaces were observed but a chaotic spot mode was found at pressures below 2kPa under conditions of high heat loss[39]. Differing mechanisms of phase formation corresponded to differing combustion modes. In the high-temperature mode, titanium nitride was present, as the initial product, at the combustion front. Solid solutions of nitrogen in titanium were alone formed in the case of the low-temperature combustion mode. For nitrogen pressures of 22 to 100kPa, a planar surface combustion mode was observed in which the reaction front propagated axially along a thin, circa ~1mm, surface layer of the sample. Reaction occurred only in this

Materials Research Forum LLC
doi: http://dx.doi.org/10.21741/9781644900093

surface layer. The maximum combustion temperature ranged from 1900 to 2200K and the mean propagation velocity ranged from 0.3 to 0.5cm/s. The deviation in the instantaneous velocity was less than 5% of the average velocity, and this mode of propagation was thus considered to be steady. Titanium nitride was the first phase to appear in the combustion front. Further nitridation of the titanium particles, with the formation of an α solid solution and further TiN_x, occurred 2 to 3s after the combustion front had passed. At nitrogen pressures of less than 2500Pa, another steady planar mode was observed, but here the reaction propagated through the bulk of the sample instead of along the surface. In this layer-by-layer regime, the reaction front moved very slowly, e.g.: 0.008cm/s at 400Pa. This mode was also associated with the relatively low combustion temperature of 1100K. The only solid solutions of nitrogen in titanium were formed in the combustion front in this low-temperature mode; titanium nitride was not detected in the post-combustion zone nor in the final product.

In so-called spin combustion there is a spiral displacement of the reaction focus[40]. It was originally thought that the various combustion regimes were characteristic of specific systems: such as self-oscillation in gas-free systems and spin in hybrid systems. Theory and experiment later showed that differing regimes could all occur during combustion in the same system, depending upon the conditions. Oscillations in reactive multilayer foils were studied by using a numerical model which took account of the effects of ambient temperature, layer roughness of the as-deposited films and local compositional inhomogeneities in the as-deposited multilayers[41].

A rough growth surface can lead to a lower effective film density if voids consequently form along columnar grain boundaries, and can render the reactant-layer thickness non-uniform. In addition, any large stresses which are present in the film can provoke decohesion from a faying surface in the form of buckling when compressive or in the form of cracking when tensile. Stresses can arise due to the ambient temperature changes, or from the effect of defects or pre-mixing at the interfaces.

It was originally presumed that a reactive multilayer having a large bilayer thickness, and thus a lesser total interfacial area between the reactants would require a higher initiation temperature because a coarse structure generally leads to a lower mixing rate at a given temperature. In fact, small-period multilayers can ignite at a lower temperature, when heated at a given rate, because the atoms need to migrate over a lower distance in order to react.

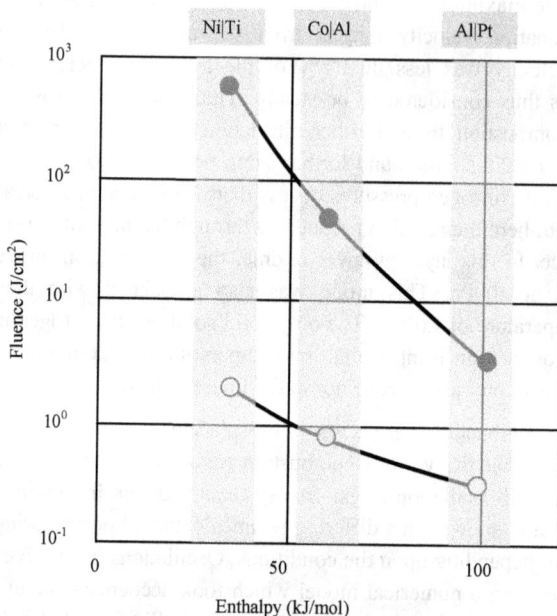

Figure 6. Ignition threshold fluence as a function of laminate enthalpy
Open circles: 108μm spot, solid circles: 8μm spot

Rather than heating an entire assembly to the ignition temperature, the reaction can be triggered by stimulating a single point in some way. This can be done by striking an electrical spark, hitting it[42], rubbing it or pointing a laser at it[43,44]. Scleroscopic investigations of mechanical initiation showed that a greater energy was required to ignite multilayers having a coarser bilayer arrangement. In a typical example of laser initiation, 5-fold multilayered Al|Ti structures, deposited onto a silicon substrate to a total thickness of 130nm, were exposed to up to 100 successive 150ps unfocused Nd:YAG laser pulses with energies of 85 or 65mJ[45]. Laser irradiation using either energy level produced almost full intermixing of the layers and the formation of intermetallic compounds. The intermixed layer|silicon interface remained intact in every case. Thinner bilayer structures have a larger number of interfaces per unit volume and, as mentioned before, the consequently greater pre-mixing markedly diminishes the amount of stored energy available for bonding. Greater energy delivery by an external source is therefore required in order to initiate reaction self-propagation in highly pre-mixed multilayers.

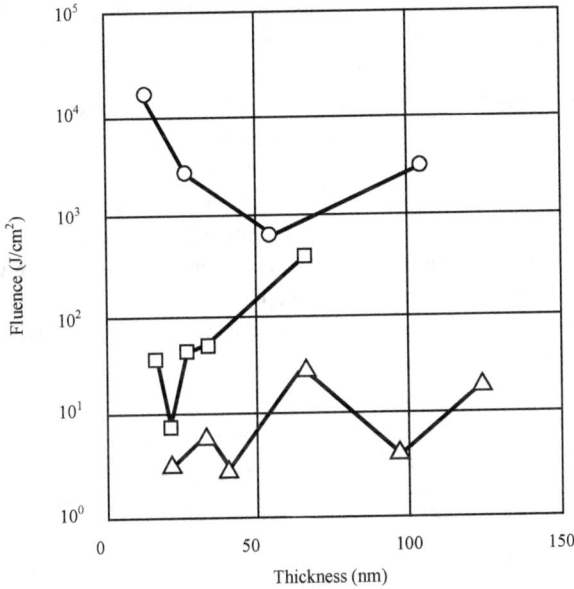

Figure 7. Ignition threshold fluence as a function of the bilayer thickness
Circles: Ni|Ti, squares: Co|Al, triangles: Al|Pt

The threshold ignition fluence also depends upon the reactive multilayer composition. In one case, the nanosecond-pulse fluences which were required to ignite Ni|Ti, Pt|Al and Co|Al foils by using a 8μm spot-size ranged over two decades. The most (-100kJ/mol) exothermic foil, Pt|Al, could be ignited by a fluence as low as some 3J/cm². The least (-34kJ/mol) energetic system, Ni|Ti required some 100J/cm². The minimum fluence also depends upon the spot size, pulse duration and wavelength. As mentioned above, there is a danger that accidental application of one of these forms of energy-input during joint set-up could prematurely initiate the reaction. The threshold levels required to ignite self-propagating reactions optically have been determined for nickel|titanium, cobalt|aluminium and aluminium|platinum nanolaminates. Lower-enthalpy pairs required higher laser-ignition fluences (figure 6). The threshold fluence for ignition by a 30ns laser pulse, with an 8μm spot-size, ranged from 0.720 to 15kJ/cm^2 for nickel|titanium, 8.6 to 380J/cm^2 for cobalt|aluminium and 3.2 to 27J/cm^2 for aluminium/platinum. Lower-enthalpy nanolaminates exhibited lower steady-state propagation rates: ranging from 0.05 to 0.9m/s for nickel/titanium, 0.6 to 8.5m/s for cobalt/aluminium and 24 to 73m/s for aluminium|platinum[46]. It was noted that energy loss via ablation may be important, especially when the spot-size is small. The ignition threshold values, for an 8μm spot size, in the case of Co|Al and Ni|Ti were well above the reported ablation thresholds (circa 10J/cm^2) for ns ultra-violet laser irradiation. Significant energy losses due to ablation may thus increase the laser fluence required for ignition. Increasing the spot size to 108μm reduced the ignition threshold fluence for all of the systems to below the problematic 10J/cm^2. The large differences in ignition fluence for the three systems, at a fixed spot diameter, were attributed to the available energy density. There was an almost reciprocal dependence of the ignition fluence (figure 7) upon the bilayer thickness as compared with the propagation speed (figure 8). For wide ranges of bilayer thickness, the ignition threshold increased with the reactant layer thickness.

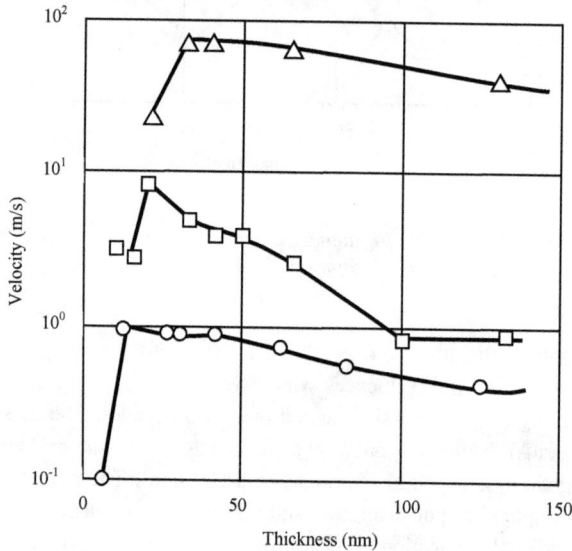

Figure 8. Reaction propagation speed as a function of the bilayer thickness Circles: Ni|Ti, squares: Co|Al, triangles: Al|Pt

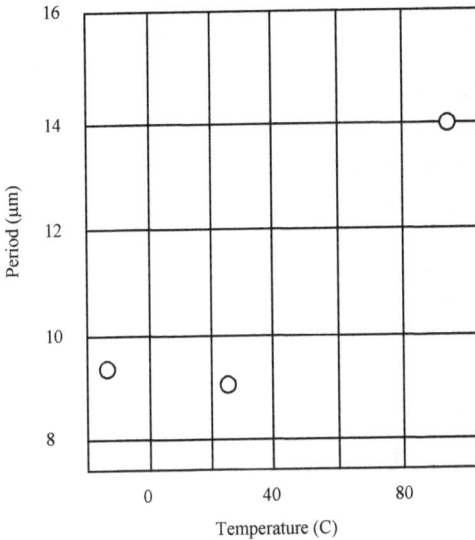

Figure 9. Surface ripple period as a function of the ambient temperature

Theoretical models and simulations can lead to an improved understanding of the processes governing steady reaction propagation in multilayers. The first theoretical models assumed that the physical properties were constant and that the layered structure was ideal[47]. The general assumption was that the two components of the layered structure would co-diffuse, react to form a single product phase and evolve the heat required to drive the reaction. Mass transport was expected to obey a single Arrhenius dependence upon temperature which was independent of the composition. Other significant variables were the volumetric heat generation rate, the heat capacity at constant pressure and the thermal diffusivity. By making the rate of heat generation proportional to the rate of concentration change, an expression could be derived the front propagation-velocity. Later models were refined by taking account of heat loss[48,49,50,51], reactant pre-mixing, phase transformation, pressure effects[52,53] and temperature-dependent properties[54,55,56]. One reduced model for simulating transient reactions in Ni|Al nanolaminates[57] relied on a simplified description of the local atomic mixing rates, which were expressed in terms of an evolution equation for a dimensionless time-scale that took account of the age of the mixed layer. The latter was in turn obtained from a theoretical analysis of a quasi one-dimensional evolution equation for a conserved scalar which yielded curves describing the mean composition and its relevant moments as a function of the local bilayer age. The

Materials Research Forum LLC
doi: http://dx.doi.org/10.21741/9781644900093

radiative heat losses were expected to be negligible when the reaction speed was high, but to become increasingly large as the front speed decreased. With regard to conductive heat loss to adjacent surfaces, experiments and modelling confirmed that the reaction was less likely to be quenched when neighboring a thick oxide layer. Such quenching was more likely to occur in contact with a conductor such as silicon.

The front propagation can also be unsteady, with large temperature variations occurring near to the reaction boundary. Auto-oscillatory one-dimensional instabilities can occur as well as two-dimensional so-called spin instabilities. The variations in microstructure which are produced by auto-oscillatory reaction fronts can affect the mechanical properties of reactive-multilayer bonds, while two-dimensional instabilities can cease suddenly at small barriers and thus lead to insufficient heat release.

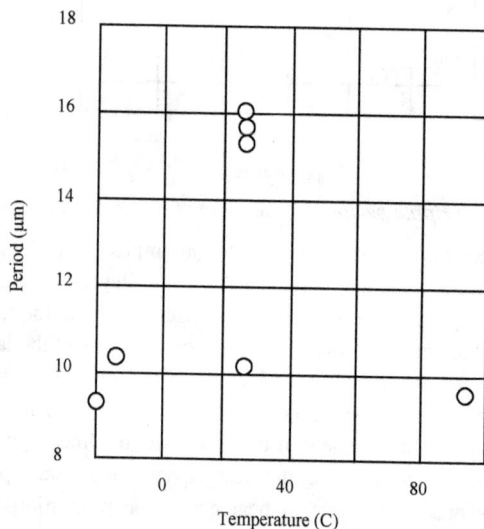

Figure 10. Band period as a function of the ambient temperature

Oscillatory reactions in planar multilayers have been predicted[58,59] on the basis of numerical and analytical models. There could be large variations in temperature, reaction-zone width, thermal width and instantaneous propagation speed. Auto-oscillation is assumed to be related to a sensitivity of the reaction to small changes in temperature: the front first moves ahead of the peak temperature but the temperature then decreases due to thermal diffusion and the peak-temperature isotherm catches up with the front, thus re-starting the cycle. The ambient temperature, interfacial roughness, compositional

Materials Research Forum LLC
doi: http://dx.doi.org/10.21741/9781644900093

heterogeneity and the degree of any pre-mixing are predicted to affect the oscillatory behaviors. Thus the energy associated with the oscillations is expected to increase with decreasing ambient temperature. Other model simulations have shown that oscillatory reaction becomes more likely with increasing pre-mixed thickness. Increasing that thickness from 1.5 to 3.5nm led to an increased magnitude and period of the oscillation. Oscillatory reaction has been reported to occur in nanostructured ZrAl||CuNi reactive multilayer foils[60]. Reactive foils were placed between copper blocks at -20 to 94C before initiating the reaction. All of the reacted foils exhibited periodic bands of coarser microstructure. The periodicity of the surface ripples and bands (figures 9 and 10) hardly changed as a function of the ambient temperature. The putative grain size in the oscillation regions was smallest for foils which were reacted at the highest ambient temperatures. No change in this grain size was observed however for foils reacted at between -20 and 25C. The variations in grain size suggested that the magnitude of the oscillations decreased when the ambient temperature increased at above room temperature. The changes in ambient temperature also had little effect upon the grain size (100nm) found in the steady-state regions (figure 11). The oscillations created periodic ripples on the foil surface and periodic variations in the length scale of the microstructure of the reacted foil which were attributed to super-adiabatic temperature oscillations at the front. The variations in the magnitude and periodicity of the oscillations agreed somewhat with predictions which were based upon numerical modelling of a different multilayer system.

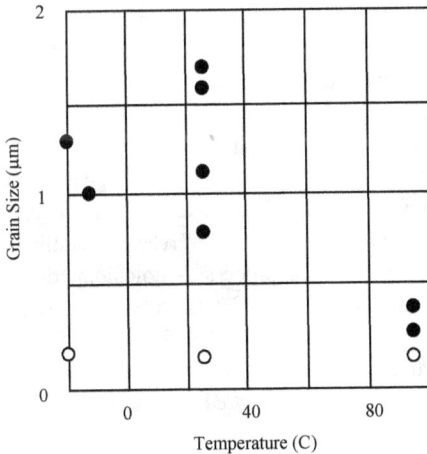

Figure 11. Grain size as a function of the ambient temperature
Open circles: steady-state region, solid circles: oscillatory region

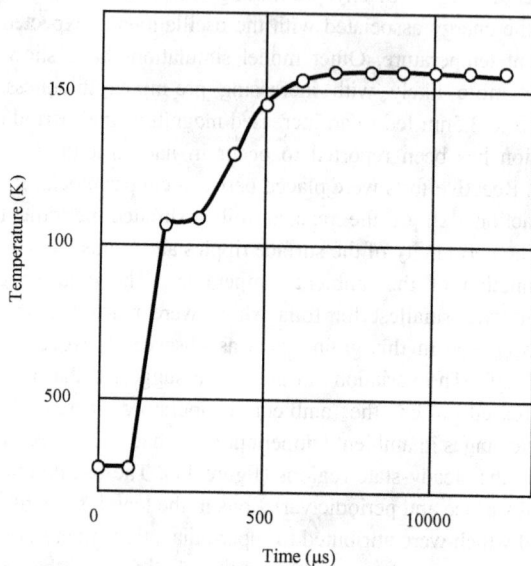

Figure 12. Thermocouple temperature-versus-time
profile of the reaction in Ni\Al multilayer foil

In general, the reaction front is smooth, travels at a constant rate and leaves behind a granular structure following re-solidification. The rate of temperature-rise which is associated with passage of the front has been estimated[61] to be about 10^6K/s. In a study of quenched exothermic waves[62], the driving mechanism of the propagation was direct exothermic dissolution of one solid reactant in a molten layer of the other. The peak temperature of the front varies from system to system but is generally much greater than 1000K. Experimental data were obtained by using an approach to the study of nanostructural transformations during the reaction of reactive multilayer foils which was based upon rapid quenching of the reaction process. A stoichiometric 1:1 Ni|Al molecular ratio multilayer foil with a bilayer thickness of 50nm and a total thickness of 15 to 20μm was studied. The reaction wave, which was locally initiated in the foil at room temperature using an electric spark propagated steadily with a velocity 9.47m/s. Temperature-time profiles (figure 12), measured by thermocouple, showed that the maximum temperature attained was 1700K; close to the melting point (1728K) of nickel. The temperature profiles indicated that the reaction was multi-stage. There was a leading stage with a very rapid (3×10^6K/s) temperature rise which required 0.5ms. This was

26

followed by a slower second stage which occupied 10ms. The NiAl reaction product had a density of 5.91g/cm^3; higher than the average density (5.17g/cm^3) of Ni|Al laminate. The bilayer spacing increased slightly within the zone of reactive dissolution; consistent with the assumption that liquid layers shrink longitudinally, expand transversally, and lead to bending of the nickel layers. A rapid decrease in the bilayer spacing ensued when the nickel layers decomposed, and NiAl grains precipitated quickly. It was concluded that the dissolution of nickel in aluminium can cause an extremely rapid self-sustained reaction, and that that became possible only because the solid product did not form a continuous layer along the boundary between the metallic layers but instead grew as separate spherical grains. This permitted nickel to dissolve in the melt throughout the whole process. It was predicted that systems (Ni|Al, Pt|Al, etc.) having a high solubility of refractory metal in the molten reactant should exhibit higher rates than would systems (Ti|Al, Zr|Al) having a low solubility.

A temperature of about 1250K has been found for 2Ni$_{91}$V$_9$|Zr, and one of 1875K for Zr|3Al[63]. The peak temperature is markedly lower when the multilayer contains a large volume fraction of pre-mixed material.

Again, the principle is best conveyed by considering common examples of the application of this bonding technique.

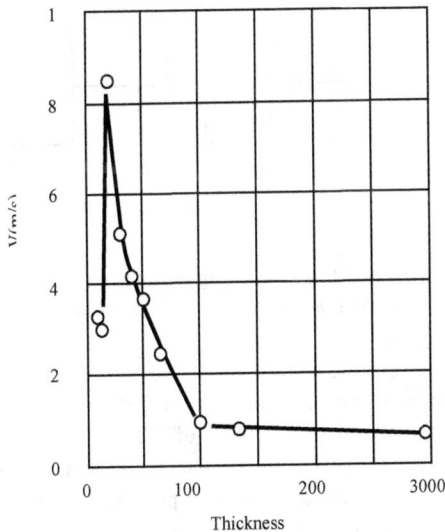

Figure 13. Propagation velocity of 7.5 μm Co|Al multilayers as a function of bilayer thickness

Materials Research Forum LLC
doi: http://dx.doi.org/10.21741/9781644900093

Al|Au

Reactive Al|Au nano-multilayered foils have been developed for potential use in the fields of dentistry and jewellery-making[64]. The reactive-foil thickness can range from 10nm to 100nm and contain many nanoscale-layers with alternating layers of material having a high heat of mixing. The reactive foil obviates the use of a furnace, and markedly increases the soldering or brazing heating-rate of the parts to be bonded. This example is interesting not only because aluminium is involved, but is not oxidized as in a thermite, but gold is also involved and that metal is usually notable for not reacting chemically with anything.

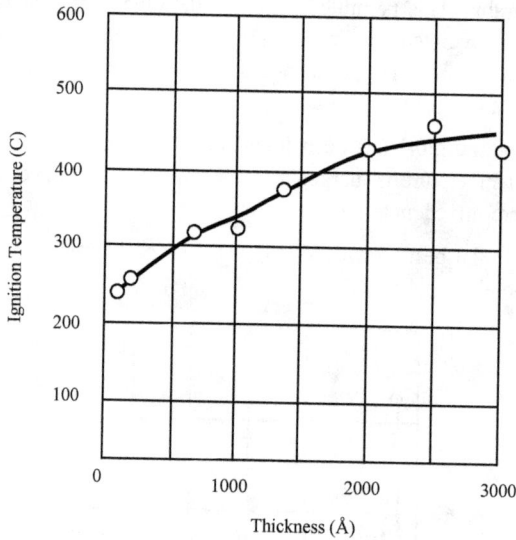

Figure 14. Ignition temperature of Co|Al multilayers as a function of bilayer thickness

Al|Co

Progress of the reaction front, stable or unstable, was studied as a function of the initial temperature, ranging from ambient to 200C, of 7.5μm-thick nanolaminated samples (figure 13). The reaction in samples with periodicities of less than 66.4nm was stable at all temperatures. The reaction in samples with periodicities greater than 100nm was unstable at all temperatures. The reaction in samples with intermediate periodicities switched from unstable to stable behavior with increasing initial temperature[65]. The

Bonding by Self-Propagating Reaction
Materials Research Foundations **45** (2019)

Materials Research Forum LLC
doi: http://dx.doi.org/10.21741/9781644900093

reaction involved both slow diffusion-limited kinetics in the region between transverse reaction bands, while more rapid kinetics occurred at their leading edges. The reaction propagated at between 0.5 and 9m/s. The foils developed various rough surface morphologies which were characterized by peak-to-valley amplitudes of 1.0μm and a multi-period wave-like structure[66]. High-temperature reaction, stimulated by rapid global heating, indicated ignition temperatures of 240 to 460C. The ignition temperature was affected by the bilayer thickness (figure 14). All of the foils which were ignited by global heating, and those which reacted via self-propagation, produced a single-phase B2 CsCl-type crystal structure. Thick nanolaminates were used to join metalized Al_2O_3.

Thermo-analytical measurements have been aimed at developing models for the ignition of pure aluminium and of aluminium-based reactive intermetallics, as well as thermites. Individual exothermic steps can be identified and attributed to particular reactions or phase transformations[67]. Various ignition stimuli have been considered, and ignition models developed which involve the detailed analysis of heat transfer, as affected by chemical reactions. There still seems to be no universally accepted ignition model. A common view is that ignition is caused by more than one reaction or phase change.

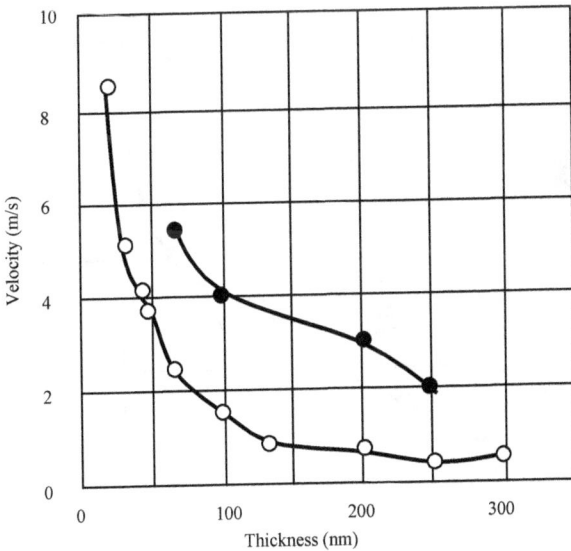

Figure 15. Reaction velocities along, and transverse to, Al\Co multilayers as a function of the bilayer thickness open circles: direct motion, solid circles: transverse motion

Materials Research Forum LLC
doi: http://dx.doi.org/10.21741/9781644900093

Vapor-deposited cobalt/aluminium nanolaminates with a net equi-atomic stoichiometry permitted rapid high-temperature synthesis. Upon being locally ignited, 7.5µm-thick cobalt/aluminium foils exhibited self-sustaining propagating reaction and the flame-front speed was between 0.5 and 9m/s. This speed was affected by the bilayer thickness and the volume fraction. The cobalt aluminide foils which resulted from the self-propagating synthesis had a rough surface morphology with a peak-to-valley amplitude of about 1.0µm and a multi-period wave-like structure. The high-temperature reaction was stimulated by rapid global heating. The ignition temperature was 240 to 460C, and was affected by the bilayer thickness. All of the foils which were ignited by global heating, and reacted in a self-propagating mode, developed a monophase CsCl-type structure[68]. The thick cobalt/aluminium nanolaminates could be used to join materials such as metalized Al_2O_3 and polyetherketone.

Those combinations which produce B2 (CsCl) structured joint materials tend to be favoured because this structure imparts moderate ductility and strength. The above forms a single B2 phase regardless of the reaction environment or reaction mode. The existence of a single product phase for Co|Al implies that the two-dimensional spin instability which is exhibited by this system has a negligible effect upon the final phase.

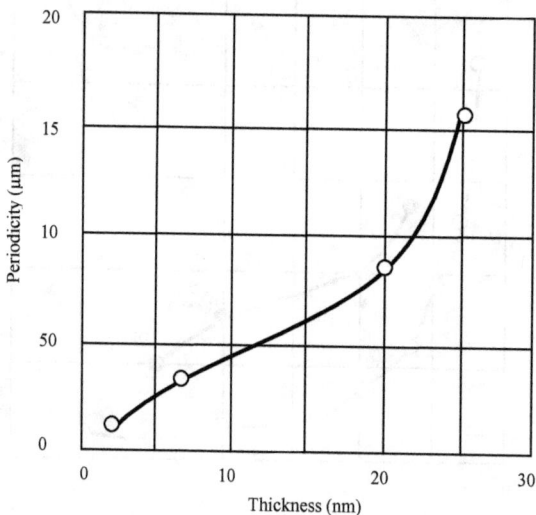

Figure 16. Surface morphology periodicity of reacted Al|Co foils as a function of bilayer thickness, deduced from scanning electron micrographs

Bonding by Self-Propagating Reaction   Materials Research Forum LLC
Materials Research Foundations **45** (2019)   doi: http://dx.doi.org/10.21741/9781644900093

Some planar multilayer systems can exhibit a two-dimensional instability which is characterized by the repeated propagation of narrow reaction bands, generally 10 to 200μm wide[69], over already-reacted parts of a multilayer. The term, spin, is applied to this instability because, in the case of cylindrical compacts, burning occurs along a spiral path. When started on one face, the reaction in bulk material propagates along a helical path while the reaction zone remains in contact with previously reacted material (figures 15 and 16). The burning spot may rotate in a single direction. In some cases, multiple reaction spots may be rotating in opposite directions. The reaction front stalls in the overall forward direction, but the material continues to react as narrow bands propagate along the warmed boundary of previously reacted material. From a limited plan viewpoint, the bands appear to move transversely to the overall forward direction. The bands which move in opposite directions exhibit a different behavior to that of those rotating in a single direction. Among the systems which exhibit two-dimensional instability are Ag|Sc, Al|Co, Al|Ni, Al|Zr, Cu|Sc and Ni|Ti.

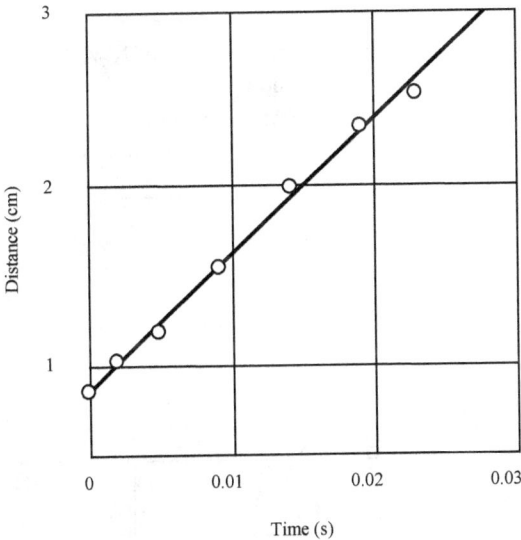

Figure 17. *Wave displacement versus distance in a 1:1 Ni|Al system*

Al|Ni

This is one of the most widely studied members of this class of bonding material, and its discussion is perhaps the best way to introduce the various themes of self-propagating reaction bonding.

Dynamic transmission electron microscopy was used to identify the resultant intermetallics and morphologies which occurred as an exothermic reaction propagated through 0.125μm-thick nanolaminated films with Al/Ni atomic ratios of 3:2, 2:3 or 1:1. Diffraction patterns with 15ns resolution indicated that NiAl alone formed, within some 15ns of the arrival of the reaction front. A transient cellular morphology occurred in Al-rich and Ni-rich foils, but not in equi-atomic ones, and this was attributed to passage through a liquid+NiAl phase-field[70]. When the Al/Ni ratio was 2:3, the propagation velocity was about 13m/s. The synthesis of Ni-Al intermetallic thin films by self-propagating reaction was again investigated for 1:1 (figure 17) and 1:3 Al/Ni stoichiometries. A sharp decrease in velocity with layer thickness (figures 18 to 20) was consistent with modelling predictions. The activation energies for NiAl synthesis ranged from 127.9 to 149.8kJ/mol, while those for the synthesis of Ni_3Al ranged from 133.8 to 146.3kJ/mol. In the former case, its magnitude was attributed to aluminium diffusion in NiAl[71]. The propagation velocity during NiAl formation ranged from 0.62 to 1.96m/s, while the propagation velocity during Ni_3Al formation ranged from 0.08 to 0.25m/s.

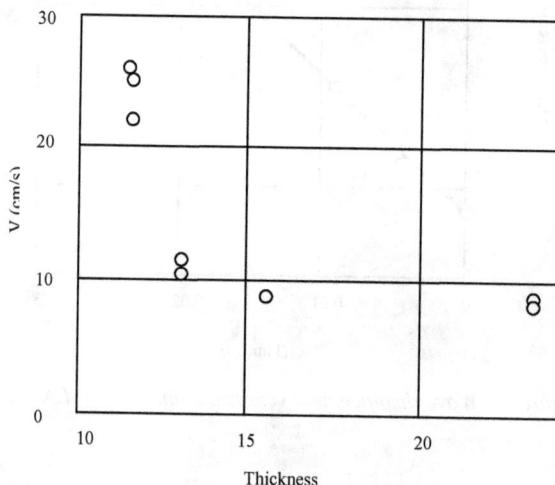

Figure 18. Wave velocity as a function of the aluminium layer thickness in a 3:1 Ni|Al system

Table 3. Propagation velocity as a function
of Al\Ni bilayer-spacing combinations

λ_1(nm)	λ_2(nm)	V(m/s)
10.4	27.1	10.4
10.4	52.1	7.5
10.5	76.7	5.0
18.7	73.6	8.8
31.5	94.5	6.5
47.7	95.5	5.5
19.1	107	6.1
45.8	204	3.5
19.1	115	12.5
45.1	191	5.1

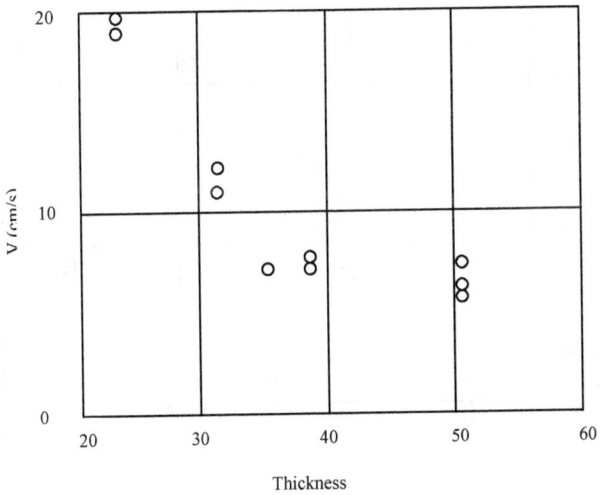

Figure 19. Wave velocity as a function of the aluminium
layer thickness in a 1:1 Ni\Al system

Materials Research Forum LLC
doi: http://dx.doi.org/10.21741/9781644900093

In another study it was interesting to note that the propagation rate was 4m/s, for Al/Ni ratios of both 1:3 and 3:1, although it seemed that melting occurred only for an atomic ratio of 3Al:1Ni[72]. Exothermic reaction was initiated in 30μm-thick multilayer foils consisting of alternating nanoscale layers of elemental metals, and occurred across a front 100μm-wide which propagated across the foil at 1 to 10m/s[73]. In Al|Ni multilayer foils, the first phases to form were Al-rich liquid and cubic intermetallic AlNi. In foils having the overall composition, AlNi, this was the stable intermetallic and was the only phase formed. In foils having the composition, Al_3Ni_2, a peritectic reaction between AlNi and the remaining liquid occurred during cooling, to form Al_3Ni_2: the stable phase at room temperature, and the final reaction product. During slow heating, there was initial formation of non-equilibrium Al_9Ni_2, with no formation of liquid phase nor AlNi. The use of Ni|3Al multilayers yields a mixture of Al_3Ni_2, Al_3Ni and Al following reaction. Subsequent annealing to 750K transforms this mixture into a single equilibrium phase, Al_3Ni. Heats of reaction and reaction velocities were determined as a function of the average bilayer spacing of sputter-deposited single-bilayer foils having a uniform bilayer spacing and of dual-bilayer foils having two (thick and thin) spacings (table 3)[74].

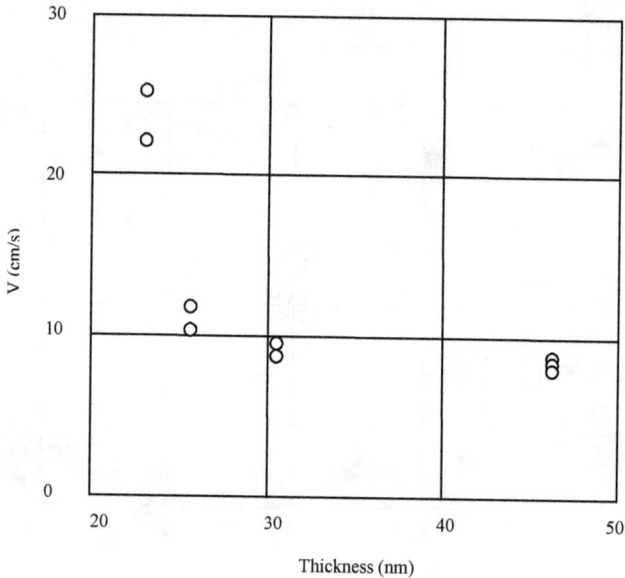

Figure 20. Wave velocity as a function of the nickel layer thickness in a 3:1 Ni|Al system

Bonding by Self-Propagating Reaction
Materials Research Foundations **45** (2019)

Materials Research Forum LLC
doi: http://dx.doi.org/10.21741/9781644900093

In the latter case, the spatial distribution of the thick and thin bilayers had a significant effect upon the reaction velocity. Coarse distributions led to much higher reaction velocities than did fine distributions. A simple model, based upon thermal diffusivities and reaction velocities could predict when the spatial arrangement of thick and thin bilayers became coarse enough to affect the reaction velocity. The propagation velocities ranged from 2.1 to 13m/s. Low-temperature annealing of sputter-deposited nanolaminate foils increased the thickness of the intermixed region between the layers. This region consisted of metastable Al_9Ni_2 while the final phase was Al_3Ni_2. Increasing the average thickness of the intermixed region from 2.4 to 18.3nm reduced the reaction velocity in all of the foils, but the effect was most marked in foils with bilayer thicknesses below 25nm. The heat-of-reaction depended upon the extent of annealing (figure 21). The reaction velocity could be divided into two distinct regimes: the first one occurred for thicker bilayers in which the average atomic diffusion distance was large. Here the reaction temperature was high, and reduction of the bilayer thickness increased the reaction velocity. The other regime occurred for thinner bilayers, where the reaction velocity was governed by the reduction in available energy, due to intermixing[75]. Reduction of the bilayer thickness here resulted in a decrease in reaction velocity.

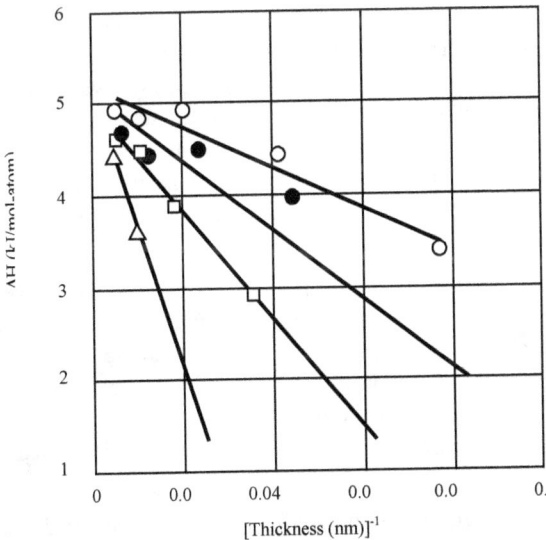

Figure 21. Heat of reaction of Ni\Ti foil as a function of the reciprocal bilayer thickness Open circles: as-deposited, solid circles: 1.5h annealing, squares: 6h annealing, triangles: 24h annealing

Materials Research Forum LLC
doi: http://dx.doi.org/10.21741/9781644900093

Overall, the propagation velocity ranged from 1.1 to 10.1m/s. This system is particularly interesting as it is the basis of a proprietary bonding product, as discussed below.

Vapor-deposited equi-atomic multilayer foils exhibit low-speed self-propagating reactions that are characterized by a spin-like reaction-front instability[76]. As well as the intermetallic reaction between nickel and titanium, reactions which take place in air can also exhibit a discrete combustion wave which is associated with titanium oxidation. In general, such an oxidation wave trails behind the intermetallic reaction front. Those multilayers which have a reactant-layer periodicity greater than 200nm exhibit a decrease in the overall reaction speed as the air pressure is reduced. Oxidation has a much smaller effect upon the overall propagation speed of multilayers having a layer periodicity of less than 100nm. This propagation speed increases when air is present, due to the additional energy which is released by titanium oxidation. The increased front speed is associated with an increased nucleation rate of the reaction bands that are typical of the spinning reaction instability of the system.

Molecular dynamics simulations, and a first-principles many-body potential, have been used to characterize the exothermic reactions of nanostructured Ni|AlNi|Al multilayers as induced by shock-loading[77]. The technique took account of the initial shock as well as of the following longer-term Ni_3Al formation. The softer aluminium layers were first shock-heated to a higher temperature than were the harder nickel layers, due to a series of reflections from the impedance-mismatched interfaces. Following initiation, the highly exothermic alloying reaction could propagate in a self-sustained manner via mass and heat diffusion. Voids played a role in the initiation of alloying, and interaction of the shock wave with the voids led to significant local heating and directly assisted the intermixing of aluminium and nickel; both making a contribution to a marked acceleration of the alloying.

Molecular dynamics simulations of Al|Ni multilayer foils similarly revealed a range of reaction pathways, depending upon the reaction temperature[78]. At the highest temperature, Fick's law interdiffusion was the rate-limiting step in intermixing. At intermediate temperatures, nickel dissolution in aluminium melt became the rate-limiting mechanism of intermixing prior to formation of the B2 intermetallic phase. At lower temperatures, the latter intermetallic formed early in the reaction process and precluded the other mechanisms.

Heats-of-reaction and heat-capacity changes were measured using scanning nanocalorimetry of nickel and aluminium bilayers at initial heating rates of 10^4K/s. Multiple exotherms were observed during the initial heating, but the number of intermediate exotherms decreased with increasing heating rate[79]. The final phase was the

Materials Research Forum LLC
doi: http://dx.doi.org/10.21741/9781644900093

B2 NiAl intermetallic. These nanocalorimeter results were compared with those from a conventional differential scanning calorimeter, operated at 0.7K/s. The high heating-rate of the nanocalorimeter delayed reaction initiation, caused the exothermic peaks to shift to higher temperatures and apparently suppressed the formation of intermediate metastable phases.

When Al|Ni reactive multilayer foils are prepared by using various methods they can have differing microstructures and exhibit different kinetic characteristics during self-propagating reaction[80]. Samples with differing structures could be prepared by using magnetron sputtering and wet etching, and differential scanning calorimetry was used to determine the effect of the ratio of the aluminium and nickel films upon the amount of heat which was generated[81].

Changes in the intermetallic phases in this reaction were studied, using a heating rate of 830K/s, in 100nm-thick bilayers which had been deposited onto a nanocalorimeter sensor[82]. Time-resolved transmission electron diffraction patterns and simultaneous thermal measurements permitted the identification of the intermetallic phases which were present. This revealed that the phase transformation sequence was essentially unaltered as compared with the effects of lower heating rates.

A molecular dynamics simulation study was made of the dissolution of nickel in liquid aluminium when using Al|Ni multilayer nanofilms, and this showed that the fundamental mechanisms underlying dissolution were diffusion-limited[83]. Subsequent intermetallic formation was explained in terms of classical nucleation theory. The results offered a better understanding of the microscopic behavior of Al|Ni reactive multilayer foils.

The effects of thermal diffusion upon the reaction-front dynamics in 1:1-ratio multilayer material have been calculated by refining an existing thermal conductivity model via the incorporation of the effects of concentration- and temperature-dependence[84]. Calculations which assumed constant and variable conductivities were compared for both axial and normal front-propagation. This revealed marked differences between the predictions; particularly with regard to the thermal and reaction widths. Differences in the average front propagation-velocity were surprisingly less marked. Some computations were performed for 3-dimensional front-propagation using constant and variable thermal conductivity models. The latter revealed the occurrence of transient spin-like reactions which appeared to be consistent with experimental observations. Stable front-propagation behavior was observed when a constant-conductivity model was used. It was concluded that thermo-diffusive instabilities probably play a role in the onset and appearance of various experimentally-observed transient front-propagation behaviours.

Materials Research Forum LLC
doi: http://dx.doi.org/10.21741/9781644900093

Processing technology for preparing reactive multilayer systems has been developed and reactive bonding has been applied to the assembly of microsystems[85]. The multilayers have been prepared by means of physical vapor deposition in the form of foils, or as direct coatings on the pieces to be bonded. The coatings can also be prepared by electroplating. The systems which have been prepared in this way include Al|Ni, Zr|Al|Si, Pd|Sn and Al|Pd. Their high thermal conductivity is useful when mounting Peltier coolers. Their thermal expansion characteristics mean that silicon-based acceleration sensors exhibit low mechanical stress after being reactive-bonded to ceramic substrates. Silicon wafers could themselves be bonded by using a Al|Ni reactive multilayer foil as a local heat source, with a heat of reaction of -57.9kJ/mol, for melting solder layers[86]. The predominant product following reaction was ordered B2 AlNi. A numerical model was developed which predicted the temperature changes in silicon wafers during bonding. This simulation revealed localized heating and rapid cooling during the reactive foil bonding[87]. Experiment showed that the joint strength of the silicon wafers was greater than the failure strength of bulk silicon.

Rotary swaging of powders into solid compacts has been considered as an inexpensive means of preparing reactive materials having refined microstructures[88]. This cold-forging reduces the diameter of tubes and, in the present case, tubes were packed with combinations of reactive powders. Diameter reduction then created an almost fully dense compact, and reduced the average reactant spacing via plastic deformation. The exothermic peaks in differential scanning calorimetry scans all shifted to lower onset temperatures with increasing degree of swaging. Meanwhile the hot-plate ignition temperature decreased and the reaction velocity increased. The shape of the initial reactants could be changed by substituting nickel flakes for nickel powders, but there was no improvement in the microstructure or the reaction behaviour. This was attributed to clumping of the nickel flakes during the initial compaction. When the aluminium powder was replaced by Al-Mg powder, the exothermic peak temperatures decreased, as did the ignition temperature, while the reaction velocity increased.

Foil samples of Al|Ni reactive multilayers could be prepared by means of hot rolling. The foils were held in a furnace before each rolling pass. Following cold rolling, and hot rolling at 200 or 300C, no reaction between the aluminium and nickel could be detected. As the rolling temperature was increased to 300C, reaction of the multilayer became harder to initiate and could be quenched before it was complete[89]. Use of higher temperatures increased the deformation during rolling and decreased the homogeneity of the foils.

The propagation velocity of the reaction front in Al|Ni multilayer films was calculated by using an extended model due to Mann which was extended by modifying the layer-thickness ratio so as to account for multilayer films having alternating layers of differing thickness[90]. The results showed that there was a critical layer thickness below which the propagation velocity increased with increasing bilayer thickness. The reverse relationship was predicted to occur above the critical value.

The various phase transformations which are involved in reaction-propagation have been studied by means of molecular dynamics; in particular, the melting of reactants, the intermixing of reactants and the formation of intermetallic compounds[91]. The reaction was found to involve two stages. The first reaction front was associated with a dissolution process and propagated at the rate of several meters per second. The second front was due to crystallization of the final product. This was slower, and led to a microstructure having alternating large grains of NiAl and molten regions in the reaction propagation direction. Grain coarsening was among the three main exothermic processes.

Stainless-steel and aluminium-alloy specimens were bonded by using nanostructured Al|Ni foils and AuSn or AgSn solder layers. For each material, a higher joining pressure increased the flow of molten solder and improved the wetting and bonding. The joint shear strength increased as the pressure which was applied during joining was increased to a critical value. At higher applied pressures, the joint shear strength remained essentially constant. The critical joining pressure depended upon the foil thickness and upon those properties of the solder which governed the duration of melting and the maximum temperature at the solder|component interface. An increasing duration and interface temperature increased the flow of solder and wetting, resulting in lower critical applied pressures.

Samples of Al|Ni multilayer foil having various bilayer thicknesses were prepared by magnetron sputtering. These foils had a stored energy of some 1100J/g, with a 4nm pre-mixed layer. As a control, copper and Al|Ni exploding foils having the same bridge size were tested using identical apparatus. The energy deposition ratio of an Al|Ni foil was 67 to 69%, while that for copper was only 39 to 45%[92].

The class of reactive materials known as inert-mediated reactive multilayers use inert materials to decouple the heating effects and maximum temperature attained. A typical such multilayer was made from 23nm bilayer (1:1-ratio) Al|Ni reactive parts and (2:3-ratio) Cu|Ni inert parts. There was negligible cross-contamination of the inert and reactive materials. There was a systematic reduction in the heat of reaction, maximum reaction temperatures and reaction propagation velocity as the volume fraction of reactive material was reduced. It was deduced that all of the samples underwent the same reaction,

Materials Research Forum LLC
doi: http://dx.doi.org/10.21741/9781644900093

Al|Ni → AlNi, but with generated temperatures ranging from 1950 to 1300K. It was further deduced that the activation energy of mixing varied markedly as the maximum reaction temperature changed. For high reaction temperatures there was a very low activation energy, of the order of 26kJ/mol; suggesting that nickel diffusion in molten aluminium was the rate-controlling mixing mechanism[93]. As the reaction temperature decreased, the activation energy tended to move to much larger values; implying that there was a change in the reaction mechanism. This was attributed to the early formation of solid products, at certain temperatures, which could impede diffusion and intermixing.

Carbon and aluminium ion-beams having various charge-states and intensities were used to irradiate sputter-deposited Al|Ni multilayer nanomaterials. Relatively short, up to 40min, irradiation markedly decreased the thermal ignition temperature and ignition delay time. Irradiation also led to atomic mixing at the Al|Ni interface, producing an amorphous interlayer with small, 2 to 3nm, Al_3Ni crystals within the amorphous material. The amorphous interlayer seemed to increase the reactivity of the multilayer nanomaterial by increasing the heat of reaction and accelerating the intermixing of nickel and aluminium. The small Al_3Ni crystals also seemed to increase the reactivity. Longer irradiation periods decreased the reactivity: causing higher ignition temperatures and longer ignition delay-times[94]. This was also attributed to the growth of Al_3Ni and to a decrease in the heat of reaction.

Foils of Al|Ni nanocomposite, having a well-known heat of reaction, were used to calibrate a non-adiabatic bomb calorimeter[95]. Reactions were initiated by using a low-energy electrical spark, and could be performed in 1atm of air, oxygen, nitrogen or argon. The bomb was designed so as to hold foil samples with minimal thermal contact, thus reducing heat losses and maximizing the surface area which was available for oxidation or nitridation. The samples were however limited to the milligram range. The calorimeter had an energy-equivalent of 279J/K, thus permitting the measurement heat outputs which were of the order of tens of joules.

A piezoelectric accelerometer having a wide dynamic range was used to monitor the vertical acceleration during the reaction of Al|Ni reactive multilayer foil. A strong signal occurred at the moment when the reaction was completed. A correlation was found between the signal height and the total volume contraction[96]. The method could be used to identify materials and processing conditions which produced much-attenuated shock-waves and thus a lower risk of damage to an assembly.

Crystal growth during self-propagating reaction in Al|Ni nanofoils was investigated using molecular dynamics simulations. Hetero-epitaxial growth of NiAl on the nickel during mixing was considered for three low-index surfaces. In the case of (001) and (111) nickel

Materials Research Forum LLC
doi: http://dx.doi.org/10.21741/9781644900093

orientations, layer-by-layer formation occurred. Four orientations of the NiAl grains were found for (001), and six for (101). Massive crystallization in the form of grains tilted with respect to the interface was observed for the (101) nickel orientation[97]. A simple geometrical construction, based upon the relationship between the unit cells of nickel and NiAl could explain the crystal growth behavior. Crystal growth varied considerably as a function of the nickel orientation.

Self-propagating reaction fronts in a network of Al‖Ni nanolayered particles were modelled numerically by using a generalized reduced continuum model which was simplified by identifying conditions under which spatial homogenisation at the particle level could be assumed. The limiting case of a single chain of particles with no porosity was analyzed for heterogeneous and homogeneous situations. A narrow region could be determined wherein the homogeneous particle approximation was valid; based upon a criterion which involved the non-dimensional ratio of the internal thermal resistance of a particle to its thermal contact resistance. The homogeneous approximation could hold true even for layered particles where the size was almost an order of magnitude greater than the thermal width of a reaction front which was steadily propagating in a uniform foil[98]. The velocity of the front in a particle compact underwent large variations as the thermal contact resistance was varied, thus offering the possibility of controlling the reaction speed and ignition behavior.

Very rapid gasless reactions in binary Al‖Ni reactive multilayer nanofoils were investigated experimentally and theoretically, with quenching being used to study the dynamics of structural transformation at the micro- and nano-scales. Experimentally obtained patterns of structural evolution for heterogeneous reactions were compared with the corresponding results for molecular dynamics simulations[99]. On the basis of the data obtained, the intrinsic mechanism of the reaction in Al‖Ni adequately explained the unusual parameters of a gasless combustion wave in such systems. Understanding reactions and structure formation in reactive multilayer nanofoils is challenging. Experimental results, based upon the combustion wave-quenching technique were here combined with molecular dynamics simulations. A comparison of experimentally observed with predicted microstructures and concentration curves revealed good qualitative and sometimes quantitative agreement. The microstructure of quenched samples included phases which existed in the combustion wave. If melt existed at the combustion temperature, it transformed into the solid phase due to rapid solidification during quenching. The NiAl$_3$ grains which were observed were attributed to the phase that formed during cooling. This phase underwent peritectic decomposition at 1127K; well below the adiabatic combustion temperature. It was perhaps possible that the maximum temperature of the self-sustained reaction, when it propagated under high heat-

Materials Research Forum LLC
doi: http://dx.doi.org/10.21741/9781644900093

loss conditions in Ni|Al foil, could be lower than 1127K. Differential scanning calorimetry, using heating rates of 0.1 to 1K/s, revealed several reaction stages; with each one corresponding to intermetallic phase formation. The intermediate phases were $NiAl_3$, Ni_2Al_9 and Ni_2Al_3; with the final reaction product being B2 NiAl. *In situ* measurements of the phase transformations occurring during the combustion of Ni|Al multilayer nanofoils, using heating-rates of 10^5 to 10^7K/s, instead indicated that the reaction occurred in a single stage and with NiAl being the only product detected. The diffusivity of nickel in liquid aluminium could explain the observed reaction rate. Solid-state diffusion was far too slow to be a major factor.

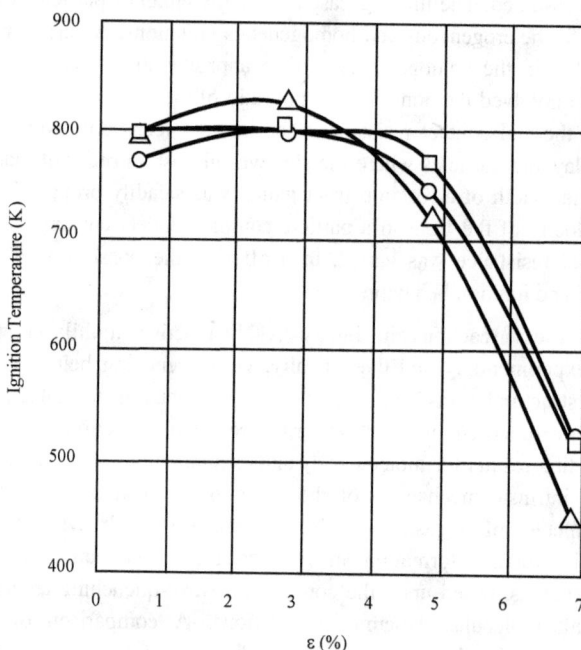

Figure 22. Predicted ignition temperatures of Al|Ni foils as a function of misfit strain Squares: 90 (111) planes, circles: 60 (111) planes, triangles: 30 (111) planes

In contrast to the usual tendency to concentrate on relatively thick Al|Ni foils for joining macroscopic parts, attention has equally been focused on the direct deposition of reactive multilayer systems[100]. Bonding of TiAl alloy was performed by depositing Al|Ni reactive multilayers, with a 14nm bilayer thickness, onto the surfaces to be joined using direct-

Materials Research Forum LLC

doi: http://dx.doi.org/10.21741/9781644900093

current magnetron sputtering[101]. The bonding was carried out at 800 or 900C while applying a pressure of 5MPa for 0.5 or 1h. Several intermetallic compounds formed at the interface and effectively bonded the TiAl. One zone of the bond comprised elongated nanograins, a second zone contained very small equiaxed grains and a third zone was characterized by larger equiaxed grains[102]. The first zone consisted of Al_2NiTi and AlNiTi while the other zones consisted of NiAl. Heat treatment at 450 or 700C transformed the Al|Ni multilayered structure into fine equiaxed NiAl grains[103]. Using foils between nanocrystalline multilayers could be a useful method for overcoming a lack of flatness of the parts to be joined, but could lead to the formation of hard brittle intermetallic compounds and a low joint shear strength[104]. In a similar study, Inconel and γ-TiAl were joined by using alternating nanolayer thin films as a filler, deposited onto each material. The nanolayers again consisted of Al|Ni exothermic reactive multilayer thin films with periods of 5 or 14nm, deposited via direct-current magnetron sputtering[105]. The latter method improved adhesion to the substrates and avoided reaction between the nickel and aluminium. Bonding was performed *in vacuo* at 800C by applying 50MPa for 1h, and was successful over large areas of the joint centre with no cracking or porosity; especially when using multilayer thin films with a 14nm modulation period.

Table 4. Misfit strain as a function of bilayer periods

m	n	ε(%)
48	56	0.58
49	56	2.68
50	56	4.77
51	56	6.87

The effects of microstructure upon reaction initiation in Al|Ni reactive multilayers were simulated by using molecular dynamics methods. Multilayer systems with various misfit strains and layer thicknesses were assumed and the ignition temperature was estimated by heating in small temperature increments until reaction occurred. The inherently incommensurate nickel and aluminium lattices were forced into various states of fractional coincidence across an interface, leading to various misfit dislocation densities and strains[106]. It was concluded that the sensitivity of reactive multilayers could be partly controlled by using the microstructure; yielding changes of the order of 350K. The magnitude of any decrease depended upon the layer thickness, with thicker bilayers exhibiting a smaller drop in ignition temperature. Nickel and aluminium layers were modelled which had [$1\bar{1}0$], [$11\bar{2}$] and [111] crystallographic directions aligned along the X, Y and Z axes, respectively. The Ni|Al interface was oriented normal to the [111] direction, and the (111) planes of atoms were stacked with a ratio of 7:3 nickel planes to aluminium planes; thus ensuring a stoichiometric ratio of about 3:1. The Ni|Al multilayer was treated as being a Z-periodic assembly of individual Ni|Al bilayers. Fractional coincidence states of 48/56, 49/56, 50/56 and 51/56 were assumed (table 4). The number of (111) planes was also be 30, 60 or 90; corresponding to reference bilayer thicknesses of 6.25, 12.5 and 18.75nm. The bilayer length and bilayer width were varied from 23.72 to 24.14nm and from 13.69 to 13.94nm, respectively. The ignition temperature was predicted to fall markedly with increasing misfit strain (figure 22)[107].

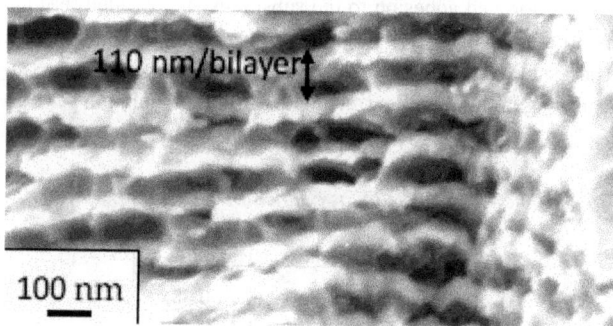

Figure 23. Scanning electron micrograph of the bilayer structure of the reactive foil, where the white areas are nickel and the grey areas are aluminium

Exothermic reaction pathways in cold-rolled Al|Ni reactive multilayer foils and physical vapor deposited Al|Ni reactive multilayer foils were investigated by heating Al|Ni samples to 1000C[108]. A series of exothermic peaks was detected using differential scanning calorimetry. The relative positions, widths and amplitudes of these peaks indicated the occurrence of stepwise reaction. Samples were heated to each peak temperature in order to identify the intermediate reaction product and the passage of the microstructure from reactant to steady-state product. The lower-temperature peaks corresponded to the formation of Al_3Ni, while the higher-temperature peaks corresponded to the conversion of Al_3Ni into AlNi[109]. The enthalpy of reaction for cold-

Materials Research Forum LLC

doi: http://dx.doi.org/10.21741/9781644900093

rolled foil was calculated to be -57.5kJ/mol; in good agreement with the -59kJ/mol formation enthalpy of AlNi. The reaction velocity during the first stage was 7mm/s for cold-rolled foils, while that for physical vapor deposited foils ranged from 1 to 30m/s[110]. The first step involved the lateral growth of Al_3Ni from isolated nucleation sites, followed by coalescence into a continuous layer. The second step involved growth of the Al_3Ni layer perpendicular to the Al|Ni interface until all of the aluminium was consumed. Because nickel was still present, the Al_3Ni could react with it and form AlNi. The reaction product in cold-rolled foil was the same as that in the physical vapor deposited foil.

The mechanical initiation of these foils can produce unpredictable results. In order to understand better the role played by mechanical properties in impact-induced ignition, the impact-responses of nanolaminate powders and high-energy ball-milled powders have been compared[111]. Both materials had almost identical thermal ignition temperatures, but the ball-milled sample had a relatively more complex microstructure in spite of the similar, 50 to 100nm, diffusion distances. The powders were mechanically loaded, using a light gas gun, within an enclosure which allowed the reactions to be monitored using high-speed imaging. Although the thermal ignition temperatures of the two powders were different by only 30C, the nanolaminate powder reacted within ms of compaction and continued for a short while before decoupling from the compaction front. The ball-milled powders reacted only after a delay of several ms and propagated as a deflagration front. It was noted that the ball-milled powders were much more ductile than were the nanolaminates, thus suggesting that the main difference between the materials during impact was the ability of the materials to fracture.

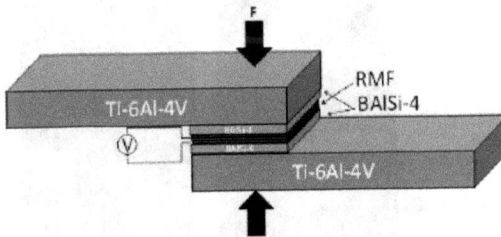

Figure 24. Test arrangement used for bonding Ti-6Al-4V by means of Al|Ni reactive multiphase foils and BAlSi-4 filler

Fracture and interparticle friction were the main mechanisms which controlled the ms reaction in nanolaminate powder. The latter was brittle, due to the presence of porosity near to the nickel grain boundaries. The ball-milled material was ductile. The nanolaminate powder reacted at the ms timescale before changing to a slower reaction mode when impacted. A prompt reaction mode occurred at 130m/s and 40MPa. The ball-milled powder reacted at the ms timescale. The estimated impact pressure was much lower than the hardness of either powder. This suggested that limited plastic deformation occurred during the initial passage of the compaction wave. The nanolaminate material exhibited brittle failure, while the other material did not. A three orders of magnitude difference in the ignition delays of the two materials was attributed to fracture of the nanolaminate powder during rearrangement. This would introduce rapid local heating and greatly increase the surface area which was subject to friction. The reaction front then travelled at the same velocity as the compaction front in nanolaminate powder. The results suggested that processes which produced a more brittle material could increase the impact-sensitivity. For low loadings, where the impact pressure was lower than the bulk strength of the powder, reaction could still occur at the ms timescale.

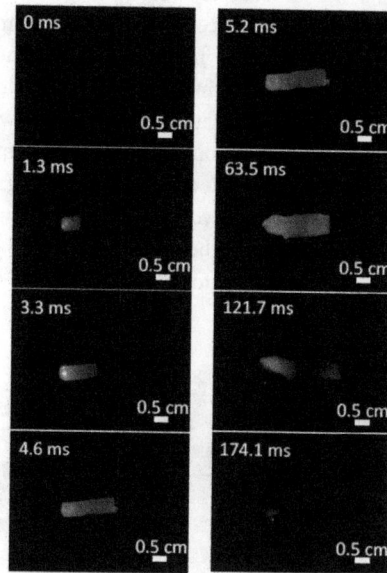

Figure 25. Progress of the reaction when bonding Ti-6Al-4V using an Al\Ni reactive multilayer film and BAlSi-4 filler

Reactions were studied in single particles of mechanically activated Al-Ni and in pellets prepared by cold pressing. The particles were ignited by using a CO_2 laser-heated hot plate at 670K. The pellets were ignited by using an electrical spark. The particles exhibited several reaction stages which corresponded to the melting of aluminium and to the peritectic decomposition of $NiAl_3$ and Ni_2Al_3. The maximum temperatures attained 1450K, before thermal losses quenched further reaction[112]. The particles within reacting pellets exhibited the same stages, but could continue to react because of the additional heat supplied by neighboring particles, and attained maximum temperatures of some 1900K. This was close to the estimated adiabatic temperature of 1911K, assuming that the final phase was NiAl. Increasing the milling time had little effect upon the temperatures and ignition delays, but increased the overall reaction rate of particles. The same sequence of phase formations occurred, regardless of the milling time, but the microstructural refinement increased the reaction rate due to diffusion-limited growth.

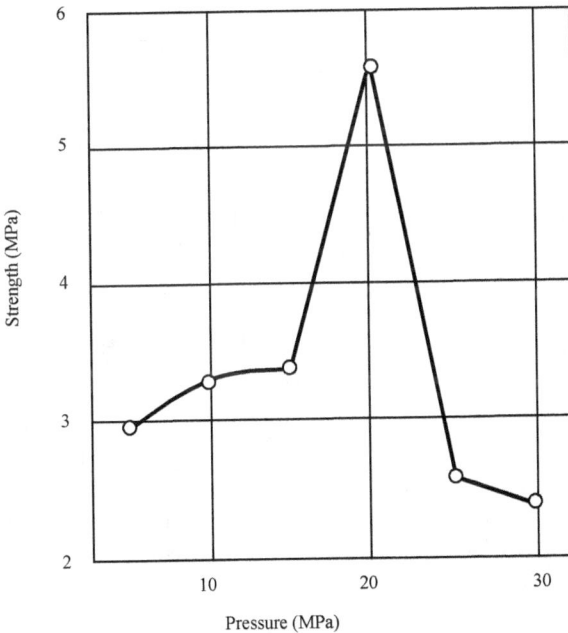

Figure 26. Bond strength of Ti-6Al-4V as a function of the applied pressure, for a total Al\Ni multilayer thickness of 66μm

Materials Research Forum LLC
doi: http://dx.doi.org/10.21741/9781644900093

Bonding of Ti-6Al-4V was carried out by using Al|Ni reactive multilayer films (figure 23), and the test arrangement shown (figure 24). This study is an excellent guide to the general pattern of behaviour followed in all reactive-multilayer joining operations. The brazing alloy, BAlSi-4, was first coated onto the Ti-6Al-4V by plasma-welding and alternating layers of nickel and aluminium up to 32.9μm thick were then coated onto the BAlSi-4 via e-beam deposition. The joint microstructure comprised AlNi and Ni$_5$Al$_3$. The presence of the two phases was attributed to differences in the diffusivities of nickel and aluminium. A maximum temperature of 683C was attained at the joint for a total foil thickness of 135μm[113]. This was easily high enough to melt the BAlSi-4, and the progress of the reaction could be followed by means of high-speed photography (figure 25). The maximum joint strength was 10.6MPa, achieved within 60s, and that strength was a function of the applied bonding pressure (figure 26) and of the total thickness of the multilayer (figure 27). Note that there was an optimum pressure. It was possible to improve the joint strength by using more ductile foils.

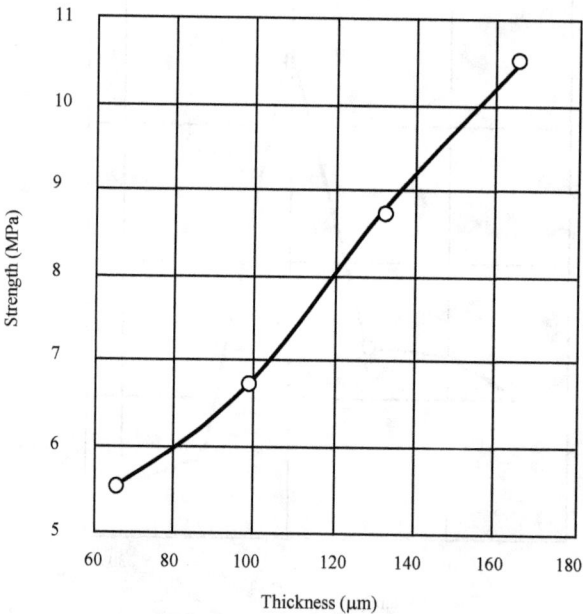

Figure 27. Bond strength of Ti-6Al-4V as a function of the total Al|Ni reactive multilayer foil thickness, for an applied pressure of 20MPa

Bonding of Ti-6Al-4V was also carried out by using electric-current heating to initiate the exothermic reaction of Al|Ni(V) multilayers and induce the diffusion of elements across the interfaces. Simulations of the temperature distribution showed that the temperature gradient in the joint was the result of differences in the resistivity of the Ti-6Al-4V alloy and the Al|Ni/Ti-6Al-4V interface. Shear testing revealed that lengthening the process from 120 to 360s improved the shear strength from 240 to 335MPa[114]. Decohesion of the joint occurred at the foil/base interface. A micro-crack network with small ridges was found on the flat surfaces of fractured samples. The resistive heating of such assemblies led to the formation of two strips of new phase at the foil|Ti-6Al-4V interface. A thinner one, which grew into the foil material, consisted of NiAl(Ti). Parts of a thicker strip, which were in contact with α-titanium grains, were filled with coarse crystalline $\alpha 2$-Ti_3Al phase. Those which were adjacent to β-titanium grains consisted of a $\alpha 2$-Ti_3Al matrix, with separate Ti_2Ni crystallites. The hardness levels of Ti-6Al-4V plates and NiAl foil material, following 360s of processing, were about 390 and 840HV, respectively[115]. These values matched those of the base materials.

Carbon-fiber/aluminium composites and TiAl alloys have been bonded via the laser-induced exothermic reaction of a Ni|Al|Ti interlayer. The addition of titanium increased the interfacial reaction between the carbon fibers and the interlayer products[116]. The formation of a 300 to 400nm thick Ti-C layer at the $NiAl_3$|carbon-fiber interface was a key factor in improving the joint quality on the carbon-fiber|Al side. The typical interfacial structure of the joint was: (C_f/Al)|Ti-C|$NiAl_3$|Ni_2Al_3|$(NiAl,Al_3NiTi_2)$|Al_3NiTi_2|TiAl. The optimum titanium content was 5wt%, as this led to a defect-free joint and a maximum shear strength of 45.8MPa.

Comparative experimental studies were made of self-sustained exothermal waves in electrodeposited amorphous antimony films, like those which inspired the present technique, in quenched glassy copper-titanium alloy and in magnetron-deposited multilayer nickel-aluminium films by using the same technique[117]. Self-propagating thermal waves were initiated in each system and converted the amorphous or multilayer structure into nanocrystalline material. The temperature-time profiles revealed a commonality of the processes in spite of their differing driving forces. Periodic microstructures which were probably caused by thermal oscillations of the crystallization fronts were observed in the antimony and CuTi. Reaction fronts in multilayer Al|Ni nanofilms, and the explosive crystallization in antimony films on thin copper substrates, were also monitored by using infra-red imaging and scanning electron microscopy. In spite of the much lower thermal effect of crystallization, the fronts of chemical reaction and crystallization were again found to have a great deal in common with regard to their temperature-time histories, including warm-up rate, temperature gradient at the front and

Materials Research Forum LLC
doi: http://dx.doi.org/10.21741/9781644900093

duration of heat release[118]. There was a micro-scintillation propagation mode of the explosive crystallization wave in antimony film.

An experimental and molecular-dynamics simulation study of combustion fronts in Al|Ni multilayer reactive nanofoils revealed previously unsuspected mechanisms. Thermal imaging and other techniques proved that the front propagation involved two stages. The first stage could propagate independently, at the same velocity as the overall front, and the products of this stage were NiAl nanograins which were separated by liquid gaps containing Al-Ni melt. Study of the intermediate and final reaction products suggested a new mechanism of so-called mosaic dissolution-precipitation that could describe nano-heterogeneous reaction during the first stage of the process[119]. This could explain the dynamics of the reaction front and the resultant microstructure. Most of the heat which was released during the second reaction stage was generated by grain coarsening. It was concluded that the reaction front in Al|Ni reactive multilayer nanofoil involved a sequential two stage process with first-stage chemical and second-stage physical exothermic transformations.

Another feature of this type of material is the ability to reverse, partially or completely, any damage which they suffer. As an example, Al|Ni(V) multilayer thin films were deposited onto tungsten wires by magnetron sputtering, and ignited by electrical discharge. Following ignition, the films underwent self-sustained reaction[120]. The as-deposited films had an irregular layered structure, with local defects of a type that was not observed on flat substrates. The as-reacted films comprised Al_3Ni_2 grains, with Al_3V at the grain boundaries. In order to use reactive multilayers for self-healing, the released heat had to be maximised by tailoring the microstructure of the nanolayered films.

Elemental aluminium and nickel powders were milled in order to determine the effect of a cryomilling atmosphere upon the microstructure and exothermic behavior. Continuous structural refinement continued for up to 8h of cryomilling, with no intermetallic compounds being detected[121]. Differential thermal analysis detected two exothermic peaks in 8h cryomilled powder, as compared with powder which was milled for 1h. The ignition temperature of the powder mixture decreased, due to gradual structural refinement. Cryomilled Al-Ni powder consisted of fine Al-Ni metastable junctions which improved the reactivity at a lower reaction temperature.

It has been shown that alloying with niobium is a means of stabilizing the as-deposited state of Al|Ni multilayers, and forming an *in situ* diffusion barrier which more than halves the heat flow-rate[122]. A comparative study was made of thermally-induced reactions in Al|Ni and Al|Ni(Nb) multilayers, confirming that such alloying was an effective strategy for tailoring the reaction kinetics of reactive multilayers.

Materials Research Forum LLC

doi: http://dx.doi.org/10.21741/9781644900093

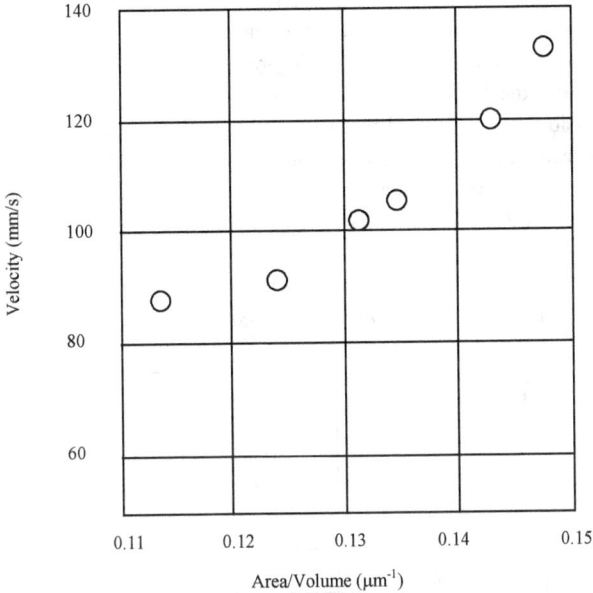

*Figure 28. Reaction velocity in Al|Ni as a function
of the contact area per unit volume*

Cold spraying has been found to be a promising technique for producing nearly fully-dense reactive bulk materials having a porosity of less than 1%. Reactive material with less than 1% of porosity and without cracking can be obtained by cold-spraying and cold-rolling, using total thickness reductions of up to 95%. A more uniform microstructure and an increased contact area between the reactant powders were produced by using more gradual rolling. In cold-sprayed materials with rolling, the onset of solid-diffusion reaction moved downwards by 30 to 40C. A clear increase in the reactivity of cold-sprayed samples, following rolling, was demonstrated. The reaction-front velocity, following fifteen gradual rolling passes, could attain 130 to 140mm/s (figure 28)[123]. This velocity exhibited a linear dependence upon the contact area per unit volume between nickel and aluminium, indicating an activation energy of 129kJ/mol.

When multilayered Al|Ni foils were heated in a highly uniform manner, ignition temperatures which were as low as 245C were found (table 5) for heating rates ranging from 2000 to 50000C/s. Four stages were involved before the reaction was complete. The first stage was that of heating to the ignition temperature. This was followed by low-temperature solid-state mixing, separate high-temperature solid-state mixing and liquid-state mixing. Various bilayer spacings, heating-rates and heating times were compared in order to show that the ignition temperature was a function of the bilayer spacing. A numerical diffusion model showed that very little chemical mixing occurred during the first 10ms of heating, whereas appreciable mixing occurred after 50ms[124]. It was suggested that the ignition temperature should increase at the lowest heating rates and that grain-boundary diffusion predominated in the early stages of solid-state diffusion.

Table 5. Ignition temperatures and incubation times for
Al|Ni nanolayer foils having a 187nm bilayer spacing

Pulse-duration (ms)	$T_{ignition}$ (C)	$t_{ignition}$ (ms)
5	270	24.4
10	245	146.6
50	290	11.2
100	320	6.2

A new approach to interpreting the differential scanning calorimetry curves of solid-state reactions with diffusion-controlled kinetics yielded an analytical expression, for the temperature at the maximum peak height, which took account of laminar, cylindrical and spherical multilayer geometries. This expression was used to analyze the Zr|CuO (table 6) and Al|Ni (table 7) multilayer nanolaminate systems. The new analysis led to geometry-independent activation energies and pre-factors. In the case of Zr|CuO, the calorimetry data scaled as the square of the bilayer thickness. In the case of Al|Ni, the calorimetry data scaled with the thickness[125]. This suggested that different reaction mechanisms operated in the thermite and aluminide systems.

The rapid energy-release by reactive multilayer foils can create high local temperature gradients across adjoining materials. An investigation was made of how differences in the properties of the constituent phases of a substrate could affect heat transport into that substrate[126]. This could then affect microstructural evolution within the substrate and thus influence the final joint properties. As an example, the effect of Al|Ni reaction upon the heat-affected zone of two Sn-Zn alloys was determined. A numerical model was developed with the aim of predicting the size of the heat-affected zone in substrate materials[127]. The model was checked experimentally by using commercially available Al|Ni multilayer foils, and Sn-Bi binary alloys.

Table 6. Peak temperatures of the first differential scanning calorimetry peak of Zr|CuO nanolaminates

Heating-rate (C/min)	Number of bilayers	T (C)
5	1	436.0
5	2	371.5
5	3	340.2
5	4	333.0
5	5	282.1
10	1	453.1
10	2	392.7
10	3	364.1
10	4	362.0
10	5	303.1
15	1	462.5
15	2	453.4
15	3	378.6
15	4	365.0
15	5	348.4
20	1	428.1
20	2	440.7
20	3	402.3
20	4	369.6
20	5	343.1
25	1	496.7
25	2	406.1
25	3	409.0
25	4	377.1
25	5	373.9

Materials Research Forum LLC
doi: http://dx.doi.org/10.21741/9781644900093

Table 7. Differential scanning calorimetry peak temperatures of Ni\Al bimetallic multilayers as a function of heating rate and bilayer thickness

Heating-rate (K/min)	Thickness (nm)	T (K)
20	20	496.9
20	40	503.2
20	80	514.4
20	160	525.8
20	320	541.0
40	20	506.8
40	40	513.6
40	80	525.6
40	160	534.7
40	320	552.4
80	20	516.4
80	40	526.3
80	80	534.5
80	160	547.0
80	320	564.5
160	20	526.1
160	40	538.8
160	80	544.9
160	160	557.6
160	320	578.2

Ceramics and copper were bonded by using reactive multilayers of nickel and aluminium which were magnetron-sputtered directly onto the ceramic, and ignited by means of electrical spark discharges or laser pulses. A high roughness of the ceramic increased the adhesion of the multilayers to the ceramic. Commercially available Nanofoil, covered with 5μm of tin on each side, were also used for comparison. Acceptable joint bond strengths were found in some cases. When using Nanofoil there was mixing of the tin coating with a silver-based thick-film metallization of the ceramic[128]. It was concluded that directly deposited reactive multilayers were promising for the bonding of ceramics to

Bonding by Self-Propagating Reaction
Materials Research Foundations **45** (2019)

Materials Research Forum LLC
doi: http://dx.doi.org/10.21741/9781644900093

other ceramics or to thermally mismatched metals, but that careful preparation of the multilayers was necessary.

It was noted[129] that previous modelling of the reactions in this system had relied upon the use of mass diffusion, and of a phenomenologically-derived diffusion coefficient to represent single-phase NiAl growth, coupled with heat transport. The reaction kinetics, temperatures and thermal-front width can however be reproduced more satisfactorily by assuming the sequential growth of Ni_2Al_3, followed by NiAl, while using independently determined interdiffusivities. A modified parabolic growth-law was used to model intermetallic growth in the thickness direction, while a multi-phase enthalpy-function was used to predict the temperature after each discrete time-step of phase growth and transformation. The results showed that Ni_2Al_3 formation produces a pre-heated zone which facilitates the slower growth of NiAl. At bilayer thicknesses of less than 12nm, the intermixing layer induces oscillation of the thermal front and markedly reduces the average velocity. The thermal-front widths, where the conversion to NiAl was complete, were estimated to be between 20 and 120μm for bilayer thicknesses of 20 to 160nm. Analytical solutions predict gradual decreases in velocity at bilayer thicknesses of less than 20nm, due to intermixing, but numerical results instead predict an abrupt transition.

Experimental observations of the magnetron-sputtering deposition of aluminium onto nickel substrates show that these sub-layers react with each other to create a new phase, especially if the nanofoil temperature is high. Such a phase unfortunately becomes a diffusion barrier to the self-propagating reaction. It is generally concluded that the diffusion barrier consists of a single phase, Al_9Ni_2, even though this phase does not exist in the relevant phase diagram. Current models also assume that diffusion-barrier formation occurs under metastable conditions, but the process is simplified by assuming isothermal solidification. It is further supposed that the barrier formed is composed of phases which are a normal feature of the Ni-Al phase diagram. Such components are either the products of peritectic transformation, preceded by the formation of primary phase or are a product solely of the solute-partitioning which results in primary-phase formation. A fundamental hypothesis is the cyclic occurrence of two phenomena. One is aluminium dissolution in the nickel substrate until some of the substrate becomes, and the other is solidification of that liquid. Barrier formation was considered only over short periods. This period has to be as short as the time required by the incubation which precedes initiation of the first solid-solid transformation.

Because aluminium is the lower melting point element in the Ni|Al nanofoil, it governs the diffusion-barrier microstructure. Thus phases/compounds possessing a high aluminium content are expected to be found in the diffusion barrier. Current models do

Materials Research Forum LLC
doi: http://dx.doi.org/10.21741/9781644900093

not exclude the formation of a zero-thickness barrier when the magnetron-deposition temperature is extremely low and a liquid zone in the substrate cannot be formed by dissolution. Diffusion-barrier formation is assumed to be a one-dimensional phenomenon. The final goal is to be able to predict the ratio of the phase thicknesses in the diffusion barrier. This is important because this ratio determines the mechanism of the self-propagating reaction in the nanofoil. It has been concluded that two phases, Al_3Ni_2 and Al_3Ni, are to be expected to be present in the diffusion barrier, and that the solidification of these phases occurs under metastable conditions. The Ni-Al phase diagram for metastable equilibrium is applicable to the description of diffusion-barrier formation, while the Ni-Al phase diagram for metastable dissolution can be used to define the optimum solute concentration in the dissolution zone. two types of diffusion occur during diffusion-barrier formation. One is boundary diffusion, which mediates dissolution, and the other is bulk diffusion which mediates solidification accompanied by peritectic transformation[130].

Numerical simulations were made of self-propagating reaction in nanoscale nickel-aluminium multilayers using Crank-Nicolson methods. The model was based upon two-dimensional heat-transfer equations, combined with heat-generation terms arising from the one-dimensional parabolic growth of Ni_2Al_3 and NiAl in the thickness direction, and with two-phase transformations such as melting and peritectic growth. The model included temperature-dependent physical and interdiffusion coefficients, specific-heat capacities and the enthalpy of reaction. The reaction-front velocity was initially controlled by the rapid growth of Ni_2Al_3, and the front temperature was limited by the peritectic reaction at 1406K. When the front had completely traversed the foil, and the temperature had reached the peritectic point, the reaction slowed down and the temperature increased due to the growth of NiAl; the only stable phase at that temperature. The reaction ceased when the initial materials had been consumed. The temperature attained the melting point of NiAl. The foil then cooled and solidified to form a final phase whose nature was governed by the overall composition.

In the case of a smaller bilayer thickness, the velocity increased due to the consequently reduced diffusion distances[131]. The relationship was reciprocal and could be fitted well using a first-order regression curve. It was noted that, in the self-propagation situation, the reaction occurred at much higher temperatures than those involved in differential scanning calorimetry. This difference might therefore modify the resultant reaction kinetics and thus make calorimetry-based predictions unreliable. In the calorimetric analysis of multilayer foils having an overall composition of Ni_2Al_3, at up to 873K, the final phase would be Ni_2Al_3 for a bilayer thickness of 50nm, but would be $NiAl_3$ plus nickel for a bilayer thickness of 12.5nm. The reaction is thus incomplete, as compared to the self-propagation situation, which occurs at above 1900K. Calorimetry which was performed at sufficiently high temperatures, such as 1407K, would indicate higher reaction enthalpies; apart from small differences due to pre-mixing and to phase formation during sputtering. The use of low-temperature (<873K) calorimetry to deduce the enthalpy of reaction was therefore untenable.

The degree of pre-mixing at very small bilayer thicknesses could be appreciable, but could not result in the elimination of front velocity: in the bilayer-thickness limit, the initial foil would be a solid solution of aluminium and nickel and would still possess excess enthalpy as compared with Ni_2Al_3 and NiAl, together with minimal diffusion distances. It could still be expected to react, albeit at a lower velocity and with a lower front temperature. The formation of intermetallic phases during sputtering could nevertheless result in zero velocity, in the thickness limit, because there was no available excess enthalpy.

Reducing front velocities with decreasing bilayer thickness can be expected as a result of increasing nanoscale effects upon heat transfer: in thin metallic layers, the thermal conductivity can decrease markedly below a layer thickness of about 50nm, due to increased electron-scattering at the interfaces and grain boundaries. This in turn can reduce the front velocity.

All the velocity curves were similar, with an initial minimum and a subsequent increase to a steady-state value. For higher Ni_2Al_3 diffusivity pre-factor values, oscillations become more significant due to increased front velocities. The average front velocity varied as the square root of the diffusivity pre-exponential factor.

When the front has completely traversed the foil, the remaining NiAl and other constituents continue to react, but at a much lower rate than that of Ni_2Al_3. This results in total reaction-times which are of the order of milliseconds. For a bilayer thickness of 40nm and a total foil length of 20μm, the total reaction time was inversely proportional to the NiAl diffusivity pre-exponential factor. The total reaction time was greater for larger foils. For a foil length of 250μm the total reaction time was 0.48ms.

The secondary reaction did not begin immediately after the passage of the front, due to heat-transfer to unreacted parts of the foil. For relatively small foil-lengths, the secondary reaction began when the first reaction was complete throughout the foil. The total reaction time therefore also depended upon the foil length, but was expected to be independent of the foil length at greater lengths.

It is clear that solid-state diffusion plays an important role in the reaction between multilayers, although relevant diffusion data should perhaps best be gathered under

Materials Research Forum LLC
doi: http://dx.doi.org/10.21741/9781644900093

conditions which are close to those which obtain in the reaction situation. So although it is interesting to know that secondary ion mass spectrometry has shown[132] that the diffusion of aluminium in nickel single crystals, between 914 and 1212K can be described by:

$$D\ (cm^2/s) = 1.0 \times 10^0\ exp[-260(kJ/mol)/RT]$$

one wonders whether the results are entirely applicable to the present situation. Rather more pertinent might be the information that the diffusion of nickel in sintered aluminium powder, when compared[133] with that in the pure bulk metal, was such that - at 46 to 540C - the rate of nickel diffusion was considerably higher than that in the bulk metal. Even quite indirect studies, but ones which better approximate the bonding situation could be useful. For example, an investigation of the aluminization of unalloyed nickel in fluoride-packs having various aluminium activities revealed that, in packs of low aluminium activity, where the Al:Ni ratio was less than 50at%, the specimen swiftly reached equilibrium. The pack remained close to equilibrium during coating. The aluminization kinetics were here controlled by diffusion in the solid. In packs of higher aluminium activity, the specimen surface did not attain equilibrium with the pack, and the process kinetics were controlled by a combination of solid and gaseous diffusion. The surface composition was time-invariant and a steady state appeared to develop at the pack|coating interface[134].

By depositing 30nm films of NiCr and 5nm films of aluminium onto FeNi substrates *in vacuo*, it was possible to determine[135] the diffusivity of nickel in aluminium (table 8).

Table 8. Diffusivity of nickel in aluminium

Temperature (K)	Diffusivity (cm^2/s)
575	2.6×10^{-14}
578	2.1×10^{-14}
585	1.5×10^{-14}
598	4.3×10^{-14}
621	1.4×10^{-13}

The volume diffusivity of nickel in aluminium was deduced[136] from the activation energy of dissociation of Al-3.6at%Ni saturated solid solutions at 150 to 300C. The activation

Materials Research Forum LLC
doi: http://dx.doi.org/10.21741/9781644900093

energy of diffusion was 2.22eV. In general, it has been shown[137] that the diffusivities of elements which have an associated enthalpy that is comparable to, or greater than, that for self-diffusion are related to the square root of the mass of the diffusant (the Varotsos-Alexopoulos rule).

Rotating-disc studies of the dissolution of nickel in liquid aluminium showed[138] that the nickel diffusivity at 700C was $6.6 \times 10^{-6} cm^2/s$, and that the diffusivity between 800 and 950C could be described by:

$$D(cm^2/s) = 1.7 \times 10^{-1} exp[-19.4(kcal/mol)/RT]$$

Other rotating-disc studies[139] of the dissolution of nickel in liquid aluminium showed that the nickel diffusivity at 700C was $1.44 \times 10^{-5} cm^2/s$, and that the diffusivity between 700 and 850C could be described by:

$$D(cm^2/s) = 5 \times 10^{-4} exp[-6.7(kcal/mol)/RT]$$

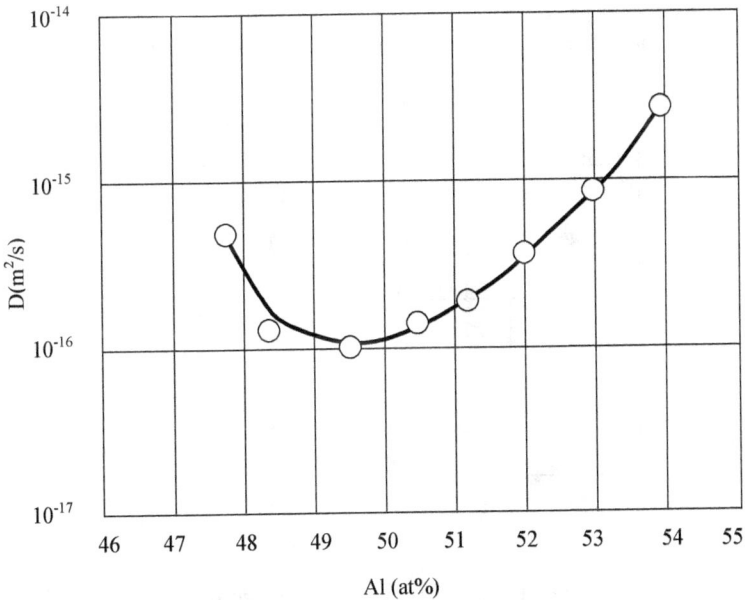

Figure 29. Interdiffusion in NiAl

Materials Research Forum LLC
doi: http://dx.doi.org/10.21741/9781644900093

Diffusion data at 950C were estimated[140] as function of concentration in multi-component β-NiAl phase grown in a Ni-based superalloy by using the pack-aluminizing technique. The calculated interdiffusion coefficients were observed to be concentration-dependent within the examined composition range (figure 29). The interdiffusion coefficients of aluminium and nickel in the multi-component β-NiAl phase were clearly lower than those reported. It was suggested that variations in the thermodynamic properties and the structural factor, due to solid-solution of the alloying elements, might be responsible for the decrease in the interdiffusion coefficients of aluminium and nickel in the multi-component β-NiAl phase.

Interdiffusivities have been deduced[141] from compositional variations which were measured across Ni|Ni₃Al diffusion couple interfaces by means of analytical electron microscopy and electron probe microanalysis (table 9). The former method was used to make measurements at lower temperatures, while the latter method was used at higher temperatures. The Boltzmann-Matano technique was used to determine interdiffusivities in the Ni-rich disordered phase and in the Ni₃Al ordered phase. These interdiffusivity values were consistent with those predicted by a modified Darken's equation.

Table 9. Interdiffusion parameters for the Ni-Al system

Al (mol%)	D_o (m^2/s)	Q (kJ/mol)
2	1.33×10^{-5}	240
4	7.50×10^{-5}	232
6	4.11×10^{-5}	248
8	5.49×10^{-5}	249
10	1.51×10^{-5}	233
12	3.43×10^{-6}	214
23	1.18×10^{-4}	277
24	1.18×10^{-4}	276

The evolution of the Ni|Al(111) interface has been studied[142] *in situ* by means of X-ray absorption spectroscopy at the Ni-K edge. Films of nickel were deposited onto bulk Al(111) to thicknesses ranging from 2 to 30 monolayers. The object being to determine the diffusion length of nickel. The nickel diffused spontaneously, at room temperature, to a depth that was estimated to be of the order of 11 monolayers. The structure of the

Ni|Al(111) mixed interface was characterized by using X-ray absorption spectroscopy. The first phase which formed on Al(111) was Al_3Ni_2-like, instead of AlNi-like. According to previous observations, an $AlNi_3$ phase formed on top of Al_3Ni_2 after deposition of the first few monolayers. It was proposed that the pure nickel growth, observed after depositing 11 monolayers, was due to the presence of $AlNi_3$; which acted as a diffusion barrier that prevented deeper nickel penetration into the aluminium at room temperature.

Interdiffusion coefficients in the B2-type phase have been determined by using single-phase diffusion couples at 1073 to 1773K. The value of the interdiffusion coefficient went through a minimum at about 47at%Al; deviating slightly from the stoichiometric composition. The value of the activation energy for interdiffusion exhibited a maximum at about 47at%Al. The magnitude of the activation energy for diffusion was suggested to be related to the lattice constant of the compound.[143] When mutual diffusion at the interface in a two-dimensional Ni–Al system was studied by means of molecular dynamics, it was established that the main diffusion mechanism consisted of correlated jumps of atoms over the vacancies near to misfit dislocation cores[144].

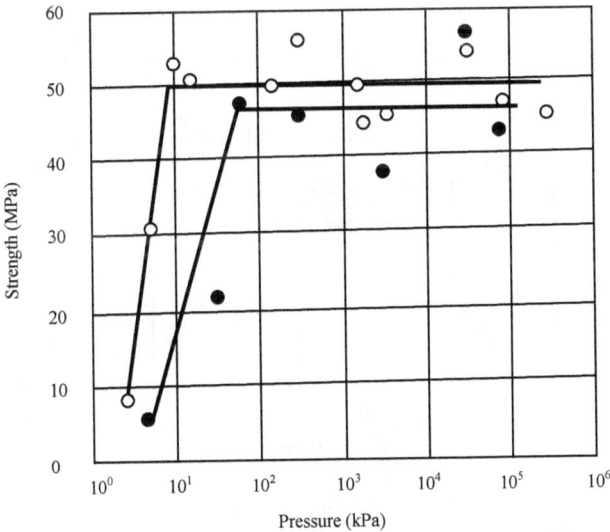

Figure 30. Tensile shear strength of stainless steel, bonded using a reactive Al\Ni foil and 25 μm AuSn solder layer, as a function of the bonding pressure Open circles: 100 μm, solid circles: 40 μm

Bonding by Self-Propagating Reaction
Materials Research Foundations **45** (2019)

Materials Research Forum LLC
doi: http://dx.doi.org/10.21741/9781644900093

The effect of a 12T high magnetic field on the intermetallic phase growth in Ni–Al diffusion couples has been studied[145]. It was found that Ni_2Al_3 and $NiAl_3$ formed in couples prepared by the pouring technique. After annealing with or without the magnetic field, the diffusion zones were still composed of Ni_2Al_3 and $NiAl_3$. The thickness of each intermetallic layer was also examined. The results demonstrated that the magnetic field reduced the thickness of Ni_2Al_3 and $NiAl_3$ layers. This reduction in intermetallic layer thickness could be attributed to the decreasing frequency factor under the magnetic field. However, the activation energy for layer growth was independent of application of the magnetic field. The magnetic field direction had an effect upon the layer growth of the Ni_2Al_3 and $NiAl_3$, which might have resulted from a change in the probability of atomic jumps in the magnetic field direction. The decrease of the layer thickness suggested that the diffusion of atoms was retarded under the magnetic field. This retardation was explained on the basis of ambipolar diffusion theory.

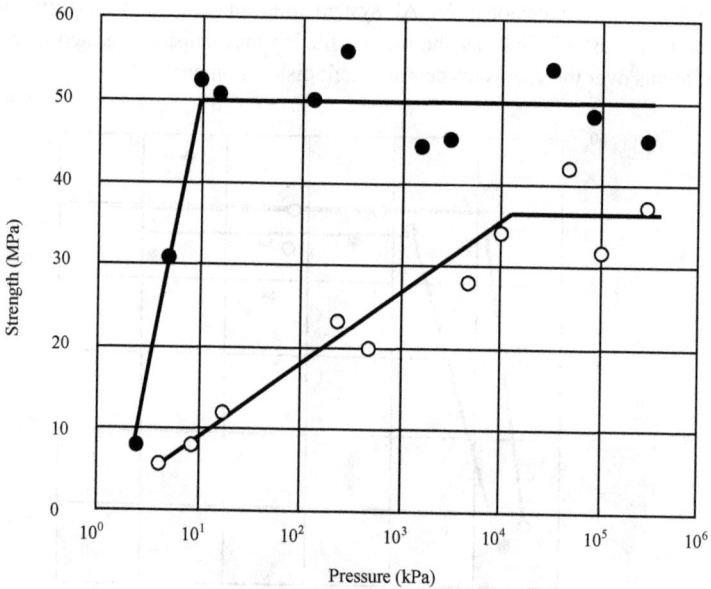

Figure 31. Tensile shear strength of stainless steel (•) and aluminium alloy (o), bonded using a 100 μm reactive Al\Ni foil and 25 μm AuSn filler layer, as a function of the bonding pressure

Bonding by Self-Propagating Reaction

Materials Research Foundations **45** (2019)

Materials Research Forum LLC

doi: http://dx.doi.org/10.21741/9781644900093

Evolution of interdiffusion microstructures was examined[146] for binary Ni–Al solid-to-solid diffusion couples using two-dimensional phase-field simulation. Utilizing semi-implicit Fourier-spectral solutions to Cahn–Hilliard and Allen–Cahn equations, multiphase diffusion couples of face-centred cubic nickel solid solution γ vs. $L1_2$ Ni_3Al solid solution γ', γ versus $\gamma+\gamma'$, $\gamma+\gamma'$ versus $\gamma+\gamma'$ with sufficient thermodynamic and kinetic database, were simulated for alloys with various compositions and volume fractions of the second phase (e.g., γ'). Chemical mobility as a function of composition was used, with a constant gradient energy coefficient, and their effect upon the final interdiffusion microstructure was examined. The microstructures were characterized by the type of boundaries formed, i.e. Type 0, Type I and Type II, in keeping with various experimental observations and thermodynamic considerations. Volume fraction profiles of the alloy phases present in the diffusion couples were measured in order to analyze quantitatively the formation or dissolution of phases across the boundaries. Kinetics of dissolution of γ' phase was found to be a function of the interdiffusion coefficients, that could vary with composition and temperature.

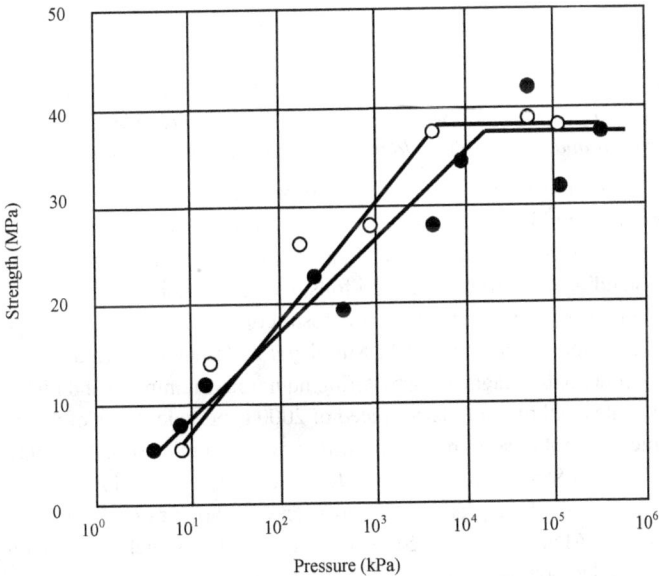

Figure 32. Tensile shear strength of aluminium alloy, bonded using a reactive Al\Ni foil and 25 μm AuSn solder layer, as a function of the bonding pressure Open circles: 160 μm foil, solid circles: 100 μm foil

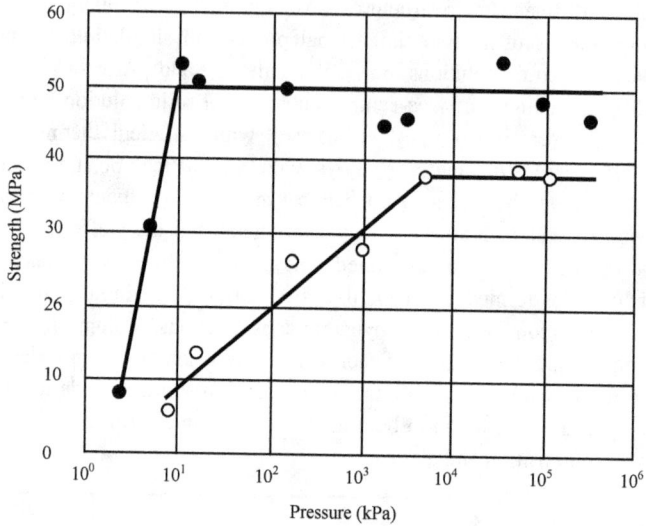

Figure 33. Tensile shear strength of stainless steel (•), bonded using a 100μm Al|Ni foil and 25μm AuSn solder layer, and of aluminium alloy (o) bonded using 160μm Al|Ni foil and 25μm AuSn solder layer, as a function of the bonding pressure

Returning to the practical application of self-propagating reaction bonding, one of the most researched multilayers is $Ni_{91}V_9|Al$; the vanadium being incorporated in order to make the material non-ferromagnetic. Stainless-steel specimens have been bonded by using free-standing nanostructured Al|Ni foils, with the reaction products, heat output and velocity being analyzed and with the microstructure and mechanical properties of the resultant joints being characterized[147]. Multilayer foils having various thicknesses were created by sequentially magnetron-sputtering numerous aluminium and nickel layers onto cooled substrates. All of the foils consisted of 2000 nanoscale layers of the two metals, in a 3-to-2 thickness ratio, so to have foils with a 50:50 atomic ratio. The bilayer thickness ranged from 35 to 85nm, with an overall foil thickness of 70 to 170μm. These foils had heats-of-reaction of 1168J/g and exhibited reaction velocities of 3.5 to 6m/s; leaving a final product of AlNi. The localized heat could completely melt free-standing sheets of AuSn filler under pressure and thus bonded the stainless steel. It was found that the tensile shear strength was 48MPa when using reactive foil joints, as compared to 38MPa for conventional bonding. As expected, both numerical methods and infra-red measurements revealed limited heat effects upon the bonded components during reactive

Materials Research Forum LLC
doi: http://dx.doi.org/10.21741/9781644900093

joining. The effect of applied pressure upon the reactive joining of stainless steel or aluminium alloy specimens using nanostructured Al|Ni foils and AuSn and AgSn solder layers was such that, for a given material, a higher applied pressure increased the flow of molten solder and thus improved wetting and bonding. The shear strengths of reactive joints increased as the applied pressure which was applied during joining increased to a critical value. When above that value, the joint shear strength remained relatively constant (figures 30 to 34). The critical pressure depended upon the foil thickness or the total heat of reaction and upon the properties of the other components, which controlled the maximum temperature at the solder|component interface[148].

Longer melting durations and higher interface temperatures enhanced melt flow, improved wetting and resulted in lower critical applied pressures. Free-standing nanostructured Al|Ni multilayer foil was similarly used to melt AuSn solder layers. When stainless-steel reactive joints were compared with aluminium reactive joints, the strengths of the stainless steel and aluminium joints increased with foil thickness[149]. The total heat of reaction increased until the foil thickness attained a critical value. When the foils were thicker than the critical value, the shear strengths were consistently of the order of 48 and 32MPa for stainless-steel joints and for aluminium joints, respectively.

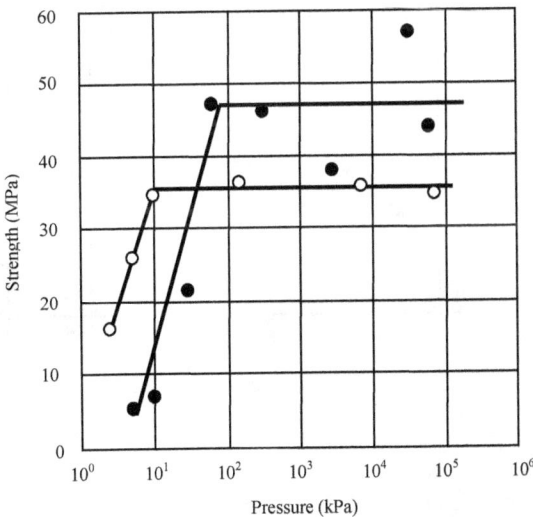

Figure 34. Tensile shear strength of stainless steel, bonded using a 40μm Al\Ni foil and 25μm filler layer (AuSn [•] or AgSn[o]), as a function of the bonding pressure

Materials Research Forum LLC
doi: http://dx.doi.org/10.21741/9781644900093

The critical foil thickness for stainless steel was 40μm, while that for aluminium was 80μm. Numerical analyses of the heat transfer during reactive joining, as well as experimental results, suggested that the melting duration of the AuSn solder was shorter when aluminium specimens were joined. A thicker foil was consequently required in order to ensure a sufficient melting duration of AuSn solder and proper wetting of the metallic samples. When components having a higher thermal conductivity, higher heat capacity and higher density were joined, the melting duration of the solder layer was shorter (figure 35). A thicker foil was therefore required in order to ensure strong joints.

The zirconium-based bulk amorphous alloy, $Zr_{57}Ti_5Cu_{20}Ni_8Al_{10}$, was bonded by using exothermic Al|Ni multilayer foils. These reactive multilayer foils were sputter-deposited at room temperature by using two separate alloy targets of Aluminum 1100 and Inconel 600. Layering was performed by rotating substrates above the two targets until the resultant bilayer thickness ranged from 40 to 69nm, with the overall thickness ranging from 87 to 274μm.

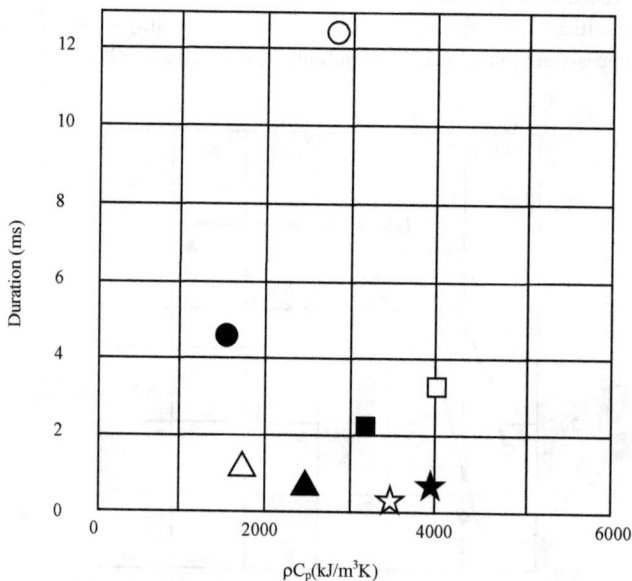

Figure 35. Numerical predictions of the melting duration as a function of the product of heat capacity, component density and thermal conductivity for a foil thickness of 80mm open circle: Ti-6Al-4V, solid circle: Pb, open square: stainless steel, solid square: Al_2O_3, open triangle: Si, solid triangle: Al, open star: Cu, solid star: Ni

Materials Research Forum LLC
doi: http://dx.doi.org/10.21741/9781644900093

The shear strength of the joints increased with the foil thickness (figure 36) and with the pressure which was applied during joining (figure 37)[150]. Shear strengths of up to 480MPa were measured using compressive single-lap specimens. As the foil reacted, and heated the surrounding metallic glass, the glass softened and flowed into cracks formed in the foil. The applied compressive load appeared to drive the softened glass into those cracks and formed continuous metallic bridges between the components, thus bonding them. Increasing the foil thickness created a greater source of heat, thus increasing the volume of material which could be heated to above the glass transition temperature of 676K and also prolonging the time during which the sample could remain sufficiently hot. As shown, increasing the applied pressure during bonding increased the force driving the softened glass into the cracks in the foil and its extrusion around the joint interface. Such a glass, unlike polycrystalline solid, exhibited a gradual decrease in viscosity over a wide range of temperatures. It was concluded that thin interfacial regions of flowing softened glass dragged the reacted foil out of the joint interface. This then widened the cracks in the foil. This implied that the viscosity of the softened glass was essential to the successful bonding of zirconium-based metallic glasses. In every case there was a sufficient flow of glass to create strong metal-metal bonds between the metallic-glass components and join them together.

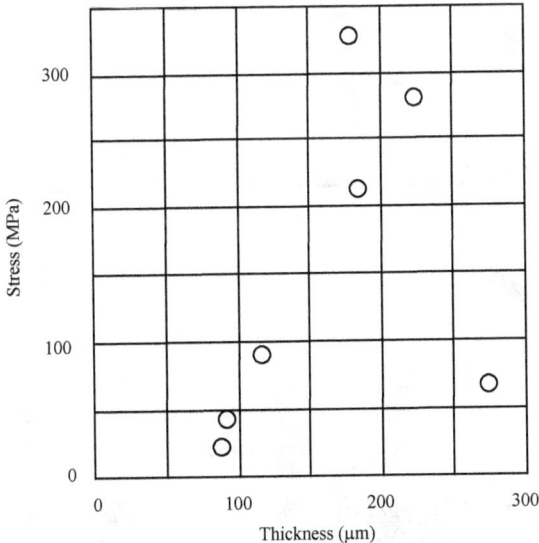

Figure 36. Compressive single-shear lap failure stress as a function of foil thickness for a joining pressure of 120MPa

Fractographic examination showed that increasing the foil thickness or the applied pressure increased the percentage of bonding along the joint interface. In the case of the strongest joints, more than 60% of the fracture surface exhibited failure of the metallic bridges between the samples, and there was a clear correlation between the percentage area of bonding, or the degree of metallic bridging, and the strength of the interface. Upon allowing for the fraction of interfacial bond area which did not contain metallic bridges, the failure stress of the joint approached the shear strength of the original metallic glass.

The fracture toughness (figure 38) of the amorphous zirconium alloy joints, made using Al|Ni reactive multilayer foils, attained a maximum value of $12MPam^{1/2}$. On the basis of the fracture toughness and crack propagation paths, it was concluded that essentially all of the toughness could be attributed to the presence of metallic glass ligaments in the joint (figure 39). Increasing the stress which was applied during joining (figure 40) increased the areal fraction of such ligaments and thus increased the fracture toughness[151].

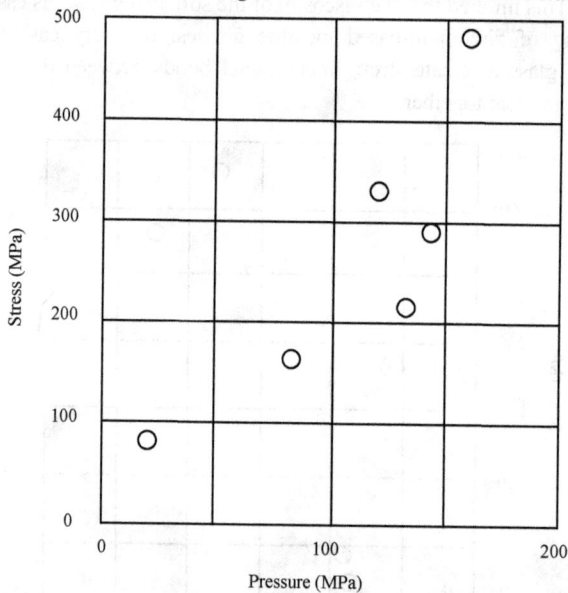

Figure 37. Compressive single-shear lap failure stress as a function of joining pressure for a foil thickness of 176μm

Materials Research Forum LLC

doi: http://dx.doi.org/10.21741/9781644900093

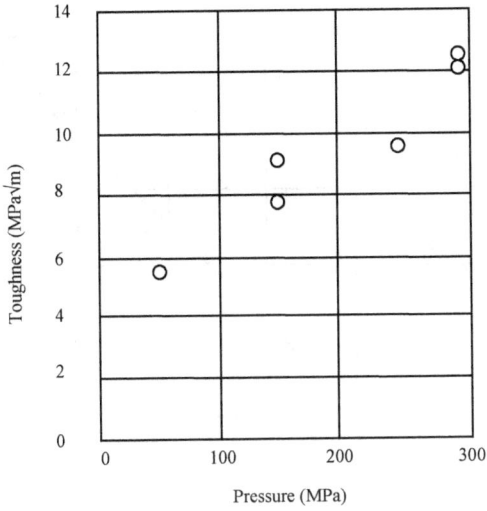

*Figure 38. Fracture toughness of bonded $Zr_{57}Ti_5Cu_{20}Ni_8Al_{10}$
as a function of the joining stress*

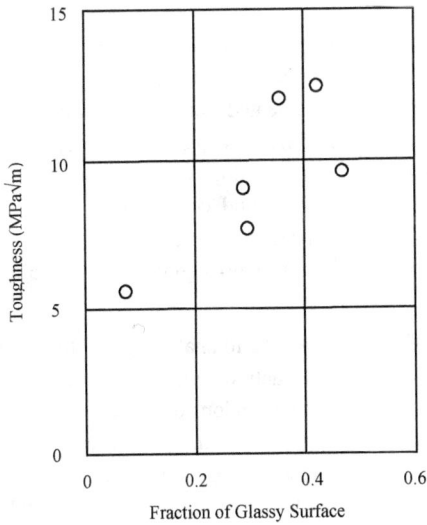

*Figure 39. Fracture toughness of $Zr_{57}Ti_5Cu_{20}Ni_8Al_{10}$
as a function of the glassy fraction of the joint*

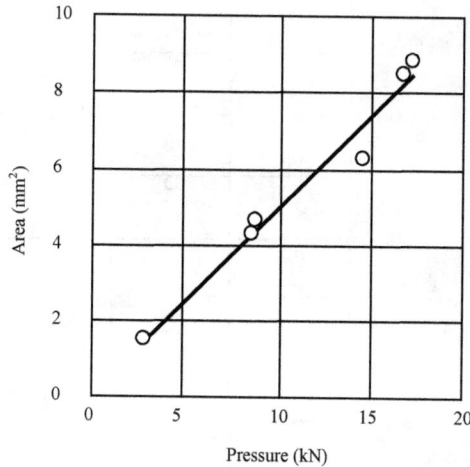

Figure 40. Area of small-scale morphology on the fracture surface of $Zr_{57}Ti_5Cu_{20}Ni_8Al_{10}$ as a function of the joining load

When the evolution of the temperature field during reactive bonding of bulk amorphous $Zr_{57}Ti_5Cu_{20}Ni_8Al_{10}$ was studied, numerical simulations predicted that the metallic glass near to the glass|foil interface was heated - at a rate of about 10^7K/s - to temperatures of some 1350K; well above the liquidus temperature (1115K) of the amorphous alloy. This was followed by rapid (10^5K/s) cooling when the reaction-front had passed. The maximum temperature, heating rate, and cooling rate of the glass all decreased with increasing distance from the interface[152]. The cooling rate here exceeded the critical cooling rate of the alloy, and there was no sign of glass crystallization resulting from the joining process.

A fluid mechanics model has been used to analyze glass fluid-flow during the bonding of bulk metallic glasses by means of reactive multilayer foils. The fracture surfaces of failed metallic glass joints exhibited distinct regions of metal-metal veining which indicated the occurrence of effective metallurgical bonding. There was a monotonic increase in the failure strength of joints when an increasing fraction of the joint comprised such veins. In the case of the strongest (420MPa shear strength) joint, nearly 60% of the fracture surface consisted of metal-metal veins[153]. It was suggested that shear stresses, due to the pressure

Materials Research Forum LLC

doi: http://dx.doi.org/10.21741/9781644900093

which was applied during joining, pushed the reactive foil away from the joint interface and create the veining.

High-resolution *in situ* X-ray diffraction analyses have been made of reactions in nanoscale foils of $Ni_{0.9}V_{0.1}|AlNi_{0.9}V_{0.1}|Al$. These data, determined to high (0.125ms) temporal and angular (0.004°) resolution over a full angular range of 120°, revealed temperature changes which corresponded to the rapid formation of NiAl, vanadium segregation and stress formation during cooling[154].

Research such as this led to the development of the commercial bonding interlayer, NanoFoil®, initially manufactured by magnetron sputter-deposition and available in thicknesses of 40 to 150μm. The product has a heat of reaction of between -44.3 and -52.2kJ/mol and can be ignited at a single point by using a laser beam or electrical impulse. The reaction then propagates at about 7m/s, leaving behind NiAl; a reasonably ductile B2-structured intermetallic. Such multilayers have been used to join bulk metallic glasses, metallized ceramics, stainless steels, titanium alloys and polymers. Bonding over areas as large as 1m², and as small as 1mm², has been achieved by using $Ni_{91}V_9|Al$ multilayers. In the latter case, even preliminary results[155] showed that the bonded interface of an infra-red emitter failed at a fracture stress of 13.5MPa, with the crack running along the solder|Covar interface and partially wandering into the silicon; thus reflecting strong bonding. Thus, it could be assumed that the bonding strength was sufficiently high. It was concluded that, in spite of the short processing time, the exothermic reaction caused the solder to melt completely and lead to homogeneous bonding. It was also significant that this bonding technique could cope with large mismatches of the coefficient of thermal expansion, such as that between steel (about 1.3 x 10^{-5}/K) and quartz (about 0.1 x 10^{-5}/K).

Initial numerical estimates of the performance of NanoFoil® indicated that it heated internally at rates of the order of 10^6K/s, while fillers and other adjacent metals naturally experienced lower rates, of the order of 10^3K/s. One analytical model[156] dealt with pre-mixing and experimental results for Al|Ni and Al|(Ni:Cu) multilayers, and concluded that pre-mixing lowers the propagation velocity by slowing the rate of atomic diffusion between layers and by lowering the reaction temperature. The lower temperature then causes solid/liquid phase changes to govern the reaction path. Depending upon the parameters chosen, propagation of the reaction front can be steady or oscillatory. Study[157] of the effect of radiative and conductive heat losses upon the propagation of reaction fronts in multilayers revealed that heat losses decreased the flame speed and the magnitude of oscillations while increasing their period. In general, radiative heat loss seemed to have a much smaller effect upon reaction propagation than did conductive heat

Materials Research Forum LLC
doi: http://dx.doi.org/10.21741/9781644900093

loss. Another simplified model[158] combined a two-dimensional diffusion equation for the atomic concentration with a quasi one-dimensional form of the energy equation which accounted for the melting of reactants and products; previous analyses having ignored melting effects. The results were applied to the evolution of self-propagating fronts in Ni|Al foils. This indicated that melting greatly affected the properties of unsteady reactions and tended to result in a significant decrease in the average speed of the front.

In a later study[159] NanoFoil®, with its alternating nickel and aluminium nanolayers and having a bilayer thickness of about 54nm with an overall thickness of 60µm, was used to bond lightweight alloys at room temperature under a pressure of between 10 and 80MPa. The nanolayers were found to be an effective method for joining titanium alloys at room temperature, and sound joints comprising mainly NiAl were obtained for TiAl|TiAl and TiAl|Inconel by using NanoFoil® and annealing at 700C for 1h under a pressure of 10MPa. A low shear strength indicated however a somewhat poor adhesion of the nanofoil to the joined materials.

Solid-state interaction between aluminium and nickel multilayer nanofilms has been investigated with regard to the phase-formation sequence during the reaction[160]. The first phase to appear was predicted on the basis of phenomenological simulation and molecular dynamics and, for the direct contact of pure nickel and pure aluminium at above 700K the first intermediate phase was expected to be a molten solution. This conclusion was based upon the metastable phase diagram, while ignoring possible intermetallic nucleation or the presence of oxides. Molecular dynamics simulation confirmed the occurrence of a decrease in the melting point of the aluminium due to dissolved nickel. When deposition was carried out at relatively high temperatures, an ordered phase with a body-centred cubic lattice was predicted to appear during deposition; thus making contact-melting impossible. When deposition occurred at room temperature, partial ordering involving two-dimensional islands in the contact zone was possible, but would not prevent contact melting.

NanoFoil® with a bilayer thickness of 50nm and a total thickness of 80µm has been subjected[161] to *in situ* transmission electron microscopic heating in order to determine the nucleation sites and growth modes of any intermetallic phases formed during the reaction. It was found that, while intermediate Al_3Ni and Al_3Ni_2 phases nucleated at Ni(V)|Al and Ni(V)|Al_3Ni interfaces, respectively, the final NiAl phase nucleated throughout the intermixed Al_3Ni and Al_3Ni_2 (originally aluminium) layer. Growth of Al_3Ni crystallites occurred as a planar front moved through the aluminium layer and converted Ni(V)|Al into Ni(V)/Al_3Ni multilayers; with the growth of Al_3Ni_2 crystallites resembling the coarsening of rounded precipitates. The final growth stage of NiAl was much faster than that of Al_3Ni_2 and resulted in the formation of a string of cubic crystallites separated by a discontinuous layer of fine Al_8V_5 crystallites.

An experiment based upon a monochromatic high-speed camera has been used[162] to monitor the shape and velocity of the propagating combustion front during the reaction, while a bomb-calorimeter was used to determine the reaction enthalpy. It was found that an empirically determined pre-exponential factor could account for the behaviour of nanofoils having various thicknesses. This was of the form,

$$T_G = 1 + (0.14.75(/\mu m)(\Gamma - \Gamma_r)$$

where Γ was the thickness of the nanofoil and Γ_r was a constant having a value of 40μm. The kinetic and energetic properties of these commercially available nanofoils could be predicted to a high degree of accuracy by using existing analytical models together with the pre-exponential factor. In a related experimental study[163], high-speed two-color infra-red pyrometry was used to monitor the temperature-profile during the reaction of nanofoils. As a result, the effective reaction temperature profile and the duration of the reaction were determined *in situ* for the first time to good temporal and spatial resolution. An analytical model could accurately predict the reaction period as a function of the layer structure of the nanofoil. The heat-affected zone around joints, produced by reactive aluminium–nickel nanofoils, was also quantified for the first time[164] by using a low melting-point alloy and a new modelling technique was developed in order to calculate the transient temperature field within the joined components during and after the exothermic nanofoil reaction. In this regard, the time spent at a high temperature depends upon the bonding technique, but is generally thought sometimes to be less than 1s.

The thickness of the filler must be such that sufficient heat is available to melt it, and also heat the adjacent material so that a good bond is obtained. The filler may flow during bonding, and even partially escape from the joint. Under an applied pressure, the filler can flow into cracks present in the reacted multilayer.

The heat released, and the phases produced, by the reaction of nickel and aluminium multilayers depend upon various details. The reaction of multilayers produced by accumulative roll bonding and sputtering is such that the heat release-rate and final NiAl grain-size are controlled by the ignition method which is used to trigger the reaction[165]. Accumulative roll bonded multilayers permit the production of joints having a greater strength than those produced using commercial sputter-produced multilayers such as NanoFoil®. On the other hand, the mechanical properties of the joint are compromised by the formation of brittle Ni_3Al, Ni_2Al_3 and $NiAl_3$.

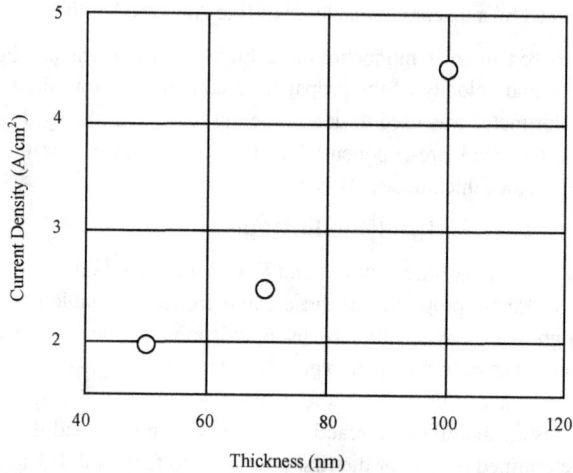

Figure 41. Current density required for the ignition of Al|Ni-7V foils as a function of the bilayer thickness

NanoFoil® has been used to melt layers between ceramics and metals[166]. By replacing the usual furnace treatment with local heating, the process avoided any significant heating of the components and areas of more than $100cm^2$ could be bonded in a relatively stress-free manner. In typical examples, Al_2O_3 was bonded to aluminium and SiC was bonded to titanium. The shear strengths were significantly higher than those obtained by using epoxy adhesives.

The ignition thresholds for provoking self-propagating reactions in Al|Ni-7V and Al|Inconel multilayer foils have been determined by using electrical (figure 41), mechanical (figure 42) and thermal (figure 43) methods. The experimental data, and theoretical models, indicated that the ignition threshold increased with increasing bilayer thickness, increasing pre-reaction intermixing and decreasing electrode contact-area. The threshold also depended strongly upon the technique which was used to ignite the multilayer. Minimizing heat loss from the ignition zone reduced the total input energy which was required for initiation, together with the ignition temperature. The electrical methods engendered high heat-losses, while hot-plate methods involved very low heat losses. The total energy-density required to ignite a 50nm bilayer Al|Ni-7V foil was 2510J/cc for a 50μs electrical pulse and a contact radius of 27μm. The equivalent energy-

Bonding by Self-Propagating Reaction
Materials Research Foundations **45** (2019)

Materials Research Forum LLC
doi: http://dx.doi.org/10.21741/9781644900093

density was only 658J/cc for ignition using a hot plate. This appreciable difference was attributed to the fact that most of the energy input was lost from the electrical-ignition test volume, before ignition, due to extensive heat loss into the surrounding foil. The higher heat losses also necessitated a higher temperature in order to initiate self-sustaining reaction in the electrical case as compared to the hot-plate case: 471C as opposed to 245C. Hot-plate experiments also yielded the most precise measurements of ignition temperature. The hot-plate ignition temperature increased from 232 to 297C as the bilayer thickness was increased from 30 to 187nm. These temperatures were much lower than the lowest (639.9C) eutectic temperature in the Al-Ni phase diagram. This showed that a self-sustaining reaction could be ignited by atomic mixing in the solid state. An activation energy of 77.3kJ/mol was calculated for this solid-state mixing; its magnitude suggesting that the diffusion of nickel into the aluminium grain boundaries predominated[167].

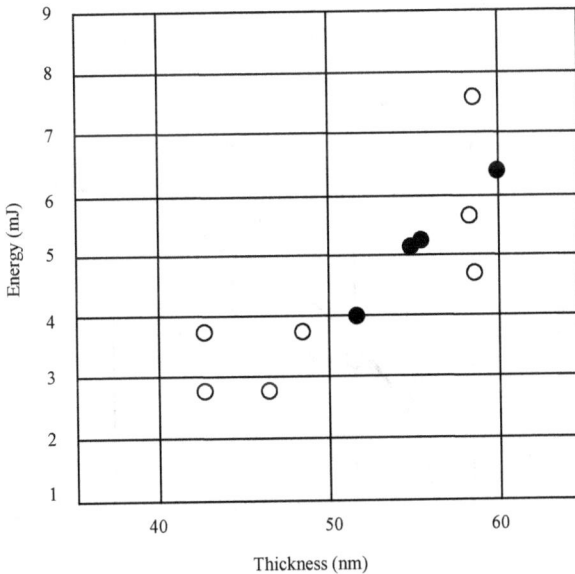

*Figure 42. Impact energy required to ignite Al\Inconel
foils as a function of the bilayer thickness
open circles: Rockwell indenter, solid circles: 1.6mm WC ball*

Materials Research Forum LLC
doi: http://dx.doi.org/10.21741/9781644900093

In practice, reactive multilayers often have to be coated with thin layers of tin, for example, before removal from a vacuum deposition system in order to prevent surface oxidation. In particular, between 0.5 and 3.0μm of the commercial product, InCuSil®, are routinely applied to NanoFoil® in order to improve wetting during reaction.

Al|Pd

Multilayers which were 1.6μm thick, with a bilayer thickness of about 180nm, were sputter-deposited onto silicon, coated with tin and pressed against another piece of silicon[168]. The heat which was generated was sufficient to melt the tin overlayer. The joint had an adhesive strength of the order of 77MPa.

Films having the required Al:Pd atomic ratio were deposited by means of alternate magnetron-sputtering from high-purity aluminium and palladium targets. The reaction velocities were up to 72.5m/s for assemblies having lateral dimensions as low as 20μm. This made feasible the bonding of silicon to silicon, or silicon to glass, at room temperature[169,170]. Shear strengths of up to 235MPa were possible without using any additional filler.

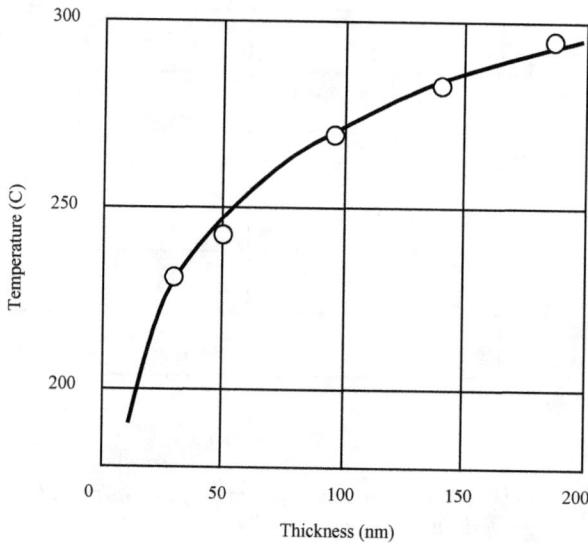

Figure 43. Hot-plate temperature required to ignite Al|Ni-7V foils as a function of the bilayer thickness

Materials Research Forum LLC
doi: http://dx.doi.org/10.21741/9781644900093

A cylindrically symmetrical finite element model for self-sustaining reactions in thin multilayer films was developed for predicting the velocity and shape of the reaction front while taking account of heat losses to a substrate, the number of bilayers and the effect of temperature-dependent specific heat capacities[171]. Multilayers of Al|Pd were used as an example. The model was able to predict the influence of heat losses and reaction properties, and the minimum number of bilayers required could be determined more precisely by including temperature-dependent specific heat capacities.

Al|Pt

These multilayers react to form a metastable rhombohedral PtAl phase. This is a complex alloy with a 76-atom unit cell[172]. The X-ray diffraction pattern of rhombohedral PtAl is unlike that of equilibrium PtAl with its cubic FeSi-type structure. Rhombohedral PtAl films transform into the equilibrium cubic phase upon slowly heating to 900K. The metastable PtAl formation was attributed to cooling rates of the order of 3.9×10^8K/s.

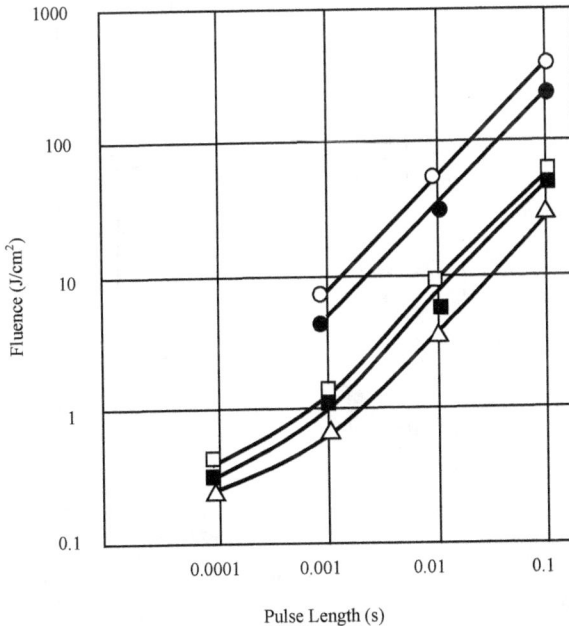

Figure 44. Ignition threshold fluence as a function of pulse length for 1.6μm-thick Al|Pt reactive foils Open circles: 102μm 65nm, solid circles: 102μm 40nm, open squares: 340μm 164nm, solid squares: 340μm 65nm, triangles: 340μm 40nm

Bonding by Self-Propagating Reaction
Materials Research Foundations 45 (2019)

Materials Research Forum LLC
doi: http://dx.doi.org/10.21741/9781644900093

Investigation of the phases which form during the solid-phase reaction of bilayer Al|Pt thin films showed that amorphous $PtAl_2$ formed first upon initiation by heating. During further heating, crystalline $PtAl_2$, Pt_2Al_3, PtAl and Pt_3Al formed sequentially[173]. This order was qualitatively consistent with an effective formation heat model. Pulsed laser irradiation could lead to single-point ignition of the rapid self-propagating reaction of multilayers having bilayer thicknesses of 328, 164 or 65nm (figure 44). For a given pulse duration, a smaller bilayer thickness required a lower laser intensity for ignition. The relationship between laser intensity and ignition onset-time could be used to deduce the activation energy for ignition[174]. The local heating rate was varied from 10^4 to 10^6K/s by adjusting the laser intensity, and the heating-rate dependence of the ignition temperature indicated an activation energy of 6.2×10^4J/mol.

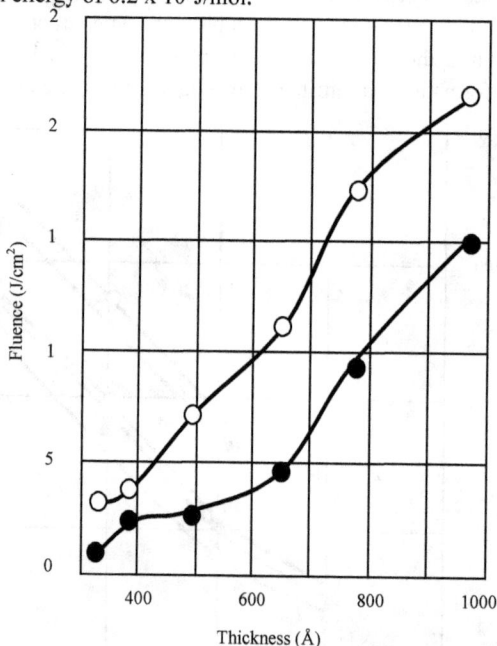

Figure 45. Femtosecond laser ignition fluence as a function of
the bilayer thickness for Al\Pt reactive multilayers open circles:
aluminium-capped, solid circles: platinum-capped

A diffusion-limited reaction model has been developed for describing Al|Pt multilayers when ignited on various substrates or free-standing[175]. Using finite-element analysis, it was used to analyze reaction-front velocity data. Both simulation and experiment indicated well-defined quench limits, as a function of the bilayer thickness, at which the heat generated by atomic diffusion was insufficient to support self-propagating reaction on a substrate. The diffusion-limited reaction model was generalized so as to allow for temperature- and composition-dependent material properties, phase changes and anisotropic thermal conductivity. This adjustment led to generally excellent agreement between simulation and experiment, although the reaction-front velocity of Al|Pt multilayers on tungsten substrates was over-estimated. Assuming a higher activation energy (47.5 instead of 41.9kJ/mol) made the results agree; with the higher activation energy being attributed to inhibited diffusion at lower heating rates. When nanolaminates with a bilayer thickness of 164nm were irradiated using single 10 or 0.5ms laser-pulses with intensities of 189 and 1189W/cm^2, respectively, shorter times-to-ignition were found for the higher power and shorter pulse combination. The irradiated area, just after ignition, exhibited a non-uniform radial brightness for longer pulses while the shorter pulses produced a uniform brightness[176]. A diffusion-limited single-step reaction mechanism was used to model the passage from reactants to products for both pulse-widths, and successfully reproduced the observed phenomena.

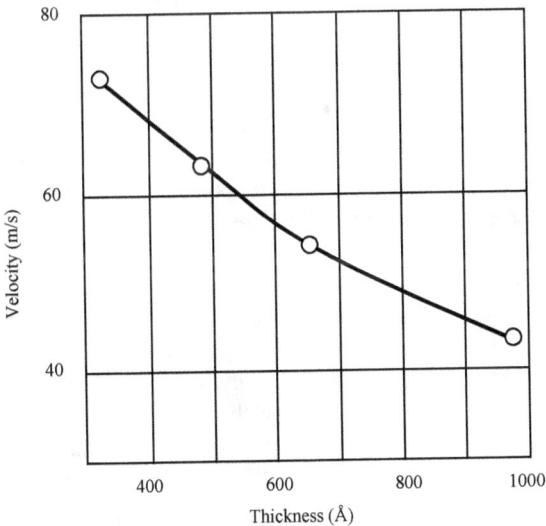

Figure 46. Reaction propagation velocity as a function of the bilayer thickness for Al|Pt reactive multilayers

Materials Research Forum LLC
doi: http://dx.doi.org/10.21741/9781644900093

Pulsed laser ignition thresholds for aluminium/platinum energetic multilayer foils have been determined as a function of the multilayer design. The results revealed differences between nanosecond and femtosecond pulsed laser-material interactions, in that the nanosecond pulsed laser ignition thresholds were lower than the femtosecond ignition thresholds; given multilayers of identical periodicity and total thickness. These differences in ignition behavior were attributed to the thermal nature of nanosecond-pulse interactions with a solid. This was different to the limited thermal diffusion and ablation of laser-heated material which was associated with femtosecond pulse interaction[177,178]. Sputter-deposited aluminium|platinum multilayer thin films exhibited rapid self-propagating high-temperature reaction. When the reactant layers were maintained at 21C before ignition and the films were deposited on oxide-passivated silicon substrates, the propagation speeds varied from some 20 to 90m/s; depending upon the bilayer dimension and the total film thickness. Contrary to the aluminium-platinum equilibrium phase diagram, the multilayers reacted in air or vacuum to give rhombohedral AlPt with a = 15.634Å and c = 5.308Å. This was stable at up to 550C, with transformation to a cubic FeSi-type structure occurring above this temperature[179]. When nanostructured aluminium/platinum multilayer films were ignited by single-pulse irradiation with a Ti:sapphire femtosecond laser, the critical fluences of 0.9 to 22J/cm^2 depended upon the multilayer design. Those with a smaller bi-layer thickness required a lower fluence, and the ignition threshold fluences were 1.4 to 3.6 times higher for aluminium-capped than for platinum-capped multilayers (figure 45). The ignition fluences converged at lower bilayer thicknesses and this was tentatively attributed to the fact that the bilayer thickness here approached the size of the optical penetration depth or that of pre-mixed AlPt interfacial layers. The propagation rate depended upon the bilayer thickness (figure 46). The capping layer seemed greatly to affect the degree of energy absorption and dissipation in the multilayer system. The bulk reflectivity was 88% for aluminium and 70% for platinum at a wavelength of 800nm. Therefore reflectivity alone could perhaps not account completely for differences in the amounts of absorbed and dissipated energy in the multilayer system. The femtosecond laser ignition of the Al|Pt multilayers was associated with ablation, so that differences in those amounts could be deduced by determining the minimum quantity of energy which was required in order to remove the top layer. To this end, monolithic 2000Å aluminium and platinum metal films were sputter-deposited onto thermally oxidized Si(100) substrates and the thin films were exposed to single pulses using fluences ranging from 0.1 to 3J/cm^2. The depth of the ablated craters for each fluence then indicated the critical fluence which was required in order to effect material removal from each metal surface. The ablation threshold fluences for aluminium (860mJ/cm^2) and platinum (540mJ/cm^2) were related to observed

Materials Research Forum LLC
doi: http://dx.doi.org/10.21741/9781644900093

differences in the ignition fluences for aluminium- and platinum-capped multilayers[180]. Given that the energy required to ablate aluminium was 1.6 times that for platinum, the ablation threshold of the capping layer was concluded to be a better explanation for differences in the amounts of energy which were dissipated into the multilayer and for the corresponding fluences which were required to ignite a self-propagating reaction.

Reactive multilayers which consisted of alternating layers of aluminium and platinum were subjected to single laser-pulses which ranged from 100μs to 100ms in duration. This resulted in the initiation of a rapid self-propagating reaction[181]. The threshold ignition intensity depended upon the focused laser-beam diameter, the bilayer thickness and the pulse length. With increasing laser pulse-length, there was a change from a more uniform to a less uniform temperature profile within the laser-heated zone. The uniform temperature profile was attributed to the rapid heating-rate and to heat localization under shorter laser pulses. A less uniform temperature profile was attributed to slower heating of the reactants and to conduction during irradiation under longer laser pulses. It was concluded that micron-scale ignition of Al|Pt occurred at temperatures which were below the melting point of either reactant.

Table 10. Diffusivity of aluminium in platinum

Temperature (C)	Diffusivity (cm^2/s)
1100	6.20 x 10^{-11}
1200	1.73 x 10^{-10}
1300	5.00 x 10^{-10}
1600	5.70 x 10^{-9}

The diffusion of aluminium at between 200 and 600C created monophase Pt_5Al_3 when annealing was performed[182] in forming-gas at 200C. At above 400C, this phase began to transform into Pt_3Al. When annealing was carried out in the presence of oxygen, the formation of Pt_5Al_3 did not occur at 200C, and only slightly at 300C. Transformation at higher temperatures was also slowed and this was attributed to the effect of oxygen in reducing the aluminium diffusivity in the platinum film.

Materials Research Forum LLC
doi: http://dx.doi.org/10.21741/9781644900093

*Table 11. Linear regression fits to the heat-of-reaction of various
Al\|Pt multilayers, where H is the heat-of-reaction (kJ/mol)
and x is the reciprocal of the bilayer thickness (nm)*

Multilayer	Equation
Al|9Pt	$H = 27.7 - 820x$
Al|4Pt	$H = 40.6 - 1010x$
Al|3Pt	$H = 53 - 628x$
Al|2Pt	$H = 76.5 - 873x$
2Al|3Pt	$H = 74.9 - 739x$
Al|Pt	$H = 93.7 - 869x$
3Al|2Pt	$H = 85.9 - 916x$
2Al|Pt	$H = 75.3 - 1110x$
3Al|Pt	$H = 61.6 - 1030x$
4Al|Pt	$H = 50.8 - 988x$
9Al|Pt	$H = 28.8 - 1120x$

Electron microprobe analysis has been used[183] to monitor the impurity diffusion of aluminium in platinum (table 10). Over this temperature range, the diffusivity could be described by:

$$D(cm^2/s) = 1.3 \times 10^{-3} exp[-193.6(kJ/mol)/RT]$$

The various rate-limiting processes which underlie ignition and reaction self-propagation in Al\|Pt multilayers have been investigated[184] by means of experiment and mathematical modelling. It was found that free-standing 1.6μm-thick multilayers ignite at temperatures which are below the melting-point of either reactant or of their eutectic; thus indicating that initiation occurs via solid-state mixing.

It was noted that equimolar multilayers required the lowest ignition temperatures, for a given bilayer thickness. An activation energy of 76.6kJ/mol was attributed to solid-state mass transport. Equimolar Al\|Pt multilayers underwent the most rapid self-sustained reactions; with wave-front velocities as high as 73m/s. Aluminium- and platinum-rich multilayers reacted at rates as low as 0.3m/s. This was consistent with a reduced heat-of-reaction and lower adiabatic temperature. Good fits were found to the experimental data (tables 11 and 12), and it was estimated that an activation energy of 51kJ/mol was associated with a mass transport which involved high heating-rates and with a thermal diffusion coefficient of $2.8 \times 10^{-6}m^2/s$ for pre-mixed interfacial volume. Platinum

Materials Research Forum LLC
doi: http://dx.doi.org/10.21741/9781644900093

dissolution in molten aluminium was a rate-limiting step in high-temperature reaction propagation in these Al|Pt multilayers.

Table 12. Linear regression fits to the ignition temperature of various Al\Pt multilayers, where T_i is the ignition temperature (C) and t is the bilayer thickness (nm)

Multilayer	Equation
Al|4Pt	$T_i = 175 + 35.9\ln[t]$
Al|3Pt	$T_i = 137.8 + 36.1\ln[t]$
Al|2Pt	$T_i = 131.8 + 36.5\ln[t]$
2Al|3Pt	$T_i = 128.9 + 33.8\ln[t]$
Al|Pt	$T_i = 103.9 + 32.1\ln[t]$
3Al|2Pt	$T_i = 123.3 + 35.2\ln[t]$
2Al|Pt	$T_i = 129.8 + 37.1\ln[t]$
3Al|Pt	$T_i = 136 + 36.7\ln[t]$
4Al|Pt	$T_i = 141.1 + 38.7\ln[t]$

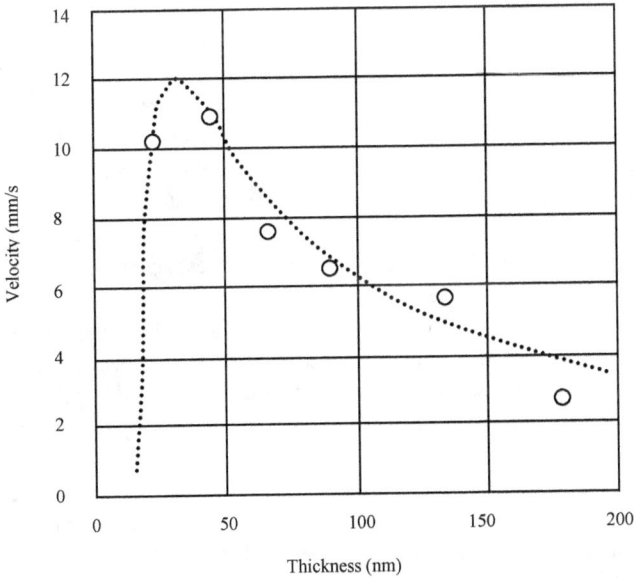

Figure 47. Instantaneous reaction front velocity as a function of Al\Ru bilayer thickness Dotted line: empirical model

Materials Research Forum LLC
doi: http://dx.doi.org/10.21741/9781644900093

Al|Ru

It was shown that this system exhibits an unusual combination of high energy-density and ductility, with reaction-front velocities of up to 10.9m/s and a peak reaction temperature of about 2000C; reflecting the high energy density[185]. The creation of a single-phase B2-RuAl microstructure ensured good ductility. A model has been developed for relating the phase transformations which occur during self-propagating reaction to macroscopic reaction parameters such as the net front velocity and the reaction temperature[186]. Coupled equations were used to describe mass and thermal transport. The temporal evolution of the temperature distribution in the reaction front was calculated as a function of the multilayer bilayer thickness. The net velocities (figure 47) were between 4.2 and 10.8m/s, with a maximum reaction temperature of up to 2171K. Interfacial pre-mixing to the extent of some 4nm had a large effect upon the reaction velocity and temperature at smaller bilayer thicknesses. The instantaneous velocity fell quickly to a minimum value and then increased again, went through a maximum and settled to a steady-state value.

Al|Ti

Multilayer foils, 15 to 20μm thick, were produced by magnetron vacuum deposition with the microstructural period ranging from 5 to 110nm and the number of layers lying between 150 and 4700. Steady-state and pulsating combustion regimes were observed. It was deduced that the most probable mechanism of self-propagating reaction was aluminium diffusion in β-titanium at a temperature which was close to that of the α → β transition[187]. For a Al:Ti ratio of unity, the propagation velocity was 0.08m/s. For a Al:Ti ratio of 3:1, the propagation rate was 0.16m/s. In another study, self-sustained reaction began at 600K and this was attributed to abnormally fast diffusion along the grain boundaries. The formation of so-called product bridges across the nanolayers was observed in the initial stages of reaction[188]. The propagation velocity depended upon the initial temperature (figure 48). The formation mechanism of the final phases, TiAl and TiAl$_3$, in the reactive multilayer films comprised several stages, beginning at 550K, and the morphology of those phases proved that the first layer appeared in column boundaries rather than along interlayer boundaries. The rate of low-temperature evolution was some four orders of magnitude lower than the rate of self-sustained reaction at high temperatures. The amount of intermediate phase produced by the low-temperature reactions therefore became negligible in the combustion regime. It was noted that single-phase well-textured TiAl formed via the gasless combustion of Ti|Al reactive multilayer foils. The rate of diffusive mixing of the reactants during the combustion reaction was 2 to 4 orders of magnitude higher than that expected on the basis of the diffusivity of aluminium in bulk solid titanium or TiAl. It was concluded that abnormally fast diffusion occurred though the inter-grain and inter-domain boundaries. Those in turn appeared to be highly disordered, or amorphous.

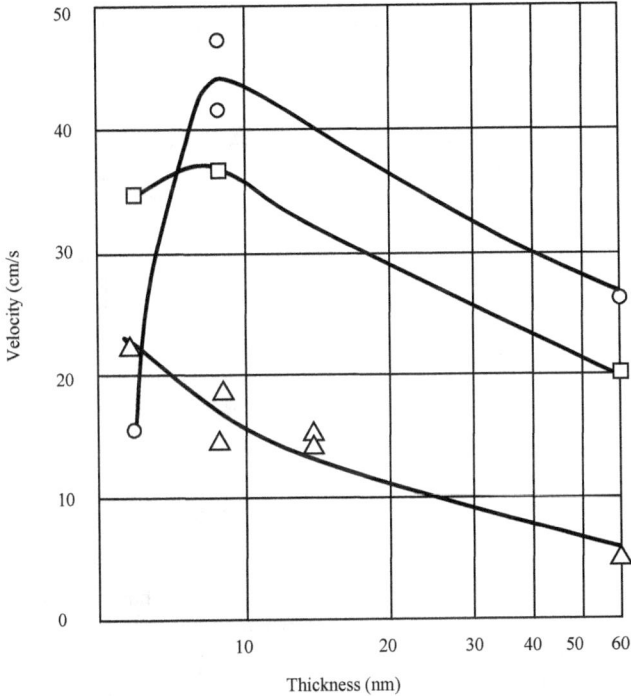

Figure 48. Reaction propagation velocity as a function of the titanium layer thickness for various initial temperatures (circles: 600K, squares: 500K, triangles: 400K)

Materials Research Forum LLC
doi: http://dx.doi.org/10.21741/9781644900093

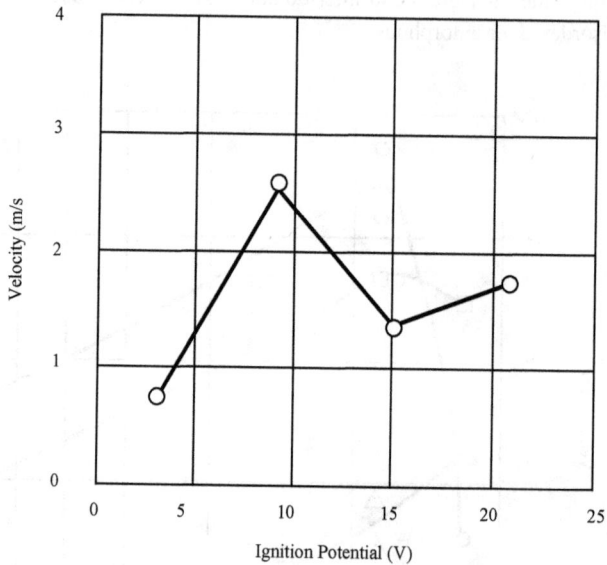

Figure 49. Reaction-front velocity versus ignition potential for 1Ti/1Al multilayer

Copper and Al_2O_3 were bonded by using Al|Ti multilayers, having various modulation periods, which had been deposited by direct-current magnetron sputtering. The bonding was performed at 900C for 600s under a pressure of 5MPa. No significant metallurgical defects were detected in the joint microstructure[189], while the existence of Cu-Ti eutectic and Al_2O_3-xTiO at the interface confirmed the success of Cu|Al_2O_3 bonding.

Reactive films having various molar ratios of titanium and aluminium, mTi|nAl (m = 1; n = 1, 2, 3), were magnetron-sputter ion-plated. The as-deposited titanium/aluminium films had columnar microstructures. Following electrical ignition, the foils underwent self-propagating reaction. Following 9V electrical ignition, the 1Ti|1Al foil had a reaction velocity of 2.6m/s; higher than that for other molar ratios. The 1Ti|1Al and 1Ti|2Al films exhibited only steady reaction propagation, while the 1Ti|3Al film exhibited both steady and unsteady propagation. The process involved both reactant mixing in the periodic layers, and grain-boundary diffusion; leading to even faster diffusion during the self-propagating reaction[190].

Bonding by Self-Propagating Reaction
Materials Research Foundations **45** (2019)

Materials Research Forum LLC
doi: http://dx.doi.org/10.21741/9781644900093

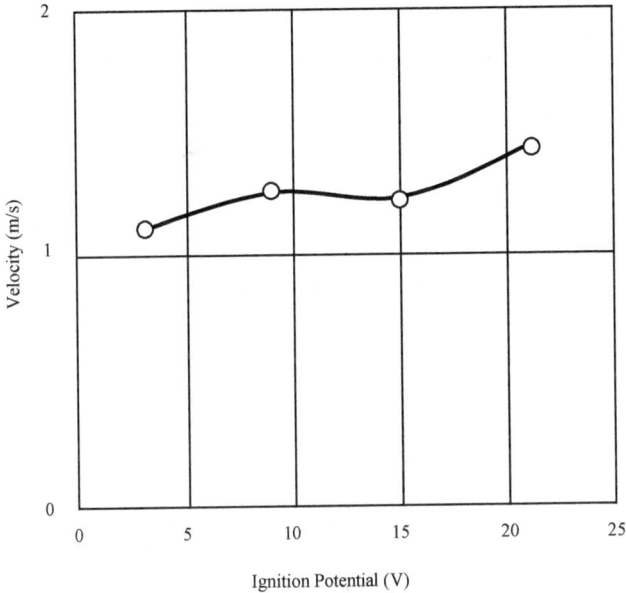

Figure 50. Reaction-front velocity versus ignition potential for 1Ti/2Al multilayer

Self-supporting Al|Ti multilayer films with various modulation structures were prepared by magnetron-sputtering and the use of a sacrificial layer. They were then ignited by means of a single pulse of radiation, yielding essentially only TiAl. The critical ignition fluences of 6 to 17J/cm² were above the typical ablation thresholds for metals under laser irradiation. The ignition fluence threshold decreased, and the propagation velocity increased, with decreasing modulation period. When the pre-mixed thickness was a large fraction of the modulation period, at smaller modulation periods, decreasing the modulation period increased the ignition threshold fluence and reduced the propagation rate[191]. For a given multilayer thickness, structures with a larger modulation period and smaller number of periods released more heat.

The reaction propagation velocity is a key parameter for characterizing multilayer foils, and a model was developed for predicting the velocity of a self-propagating reaction in multilayer foils having alternating layers of differing thickness. The reaction velocities of

Materials Research Forum LLC
doi: http://dx.doi.org/10.21741/9781644900093

Al|Ti multilayer foils were calculated by using this model[192]. There was a dependence of the critical layer thickness upon the pre-mixing thickness.

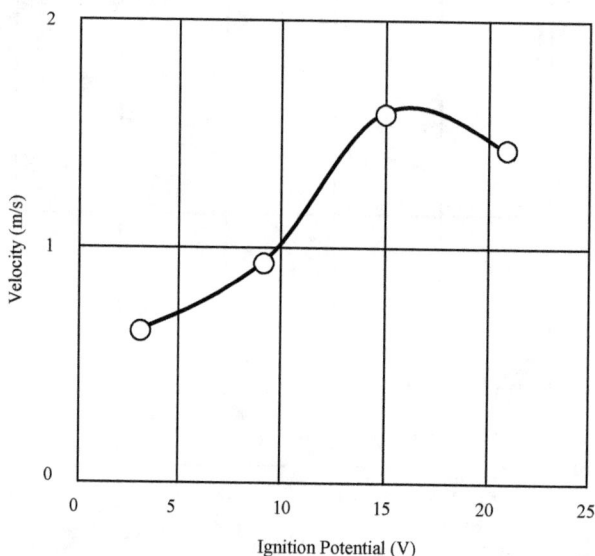

Figure 51. Reaction-front velocity versus ignition potential for 1Ti/3Al multilayer

The interaction of ultra-short laser pulses with ion-sputtered nanolayered thin film was investigated. Single-pulse irradiation was carried out in ambient air using focused and linearly polarized femtosecond pulses[193]. At low pulse-energy/fluences, ablation of only the upper titanium layer occurred. At higher fluences, there was two-step ablation of both the titanium and aluminium layers, followed by partial removal of the nanolayer film.

When magnetron-sputtered foils of the form, 1Ti|1Al, 1Ti|2Al or 1Ti|3Al, were electrically ignited, the reaction front propagated with a velocity of between 68 and 2.57m/s, depending upon the ignition potential (table 10) and the chemical composition (figures 49 to 51)[194]. Without pre-heating, all of the foils exhibited steady-state reaction propagation following electrical ignition. At lower ignition potentials, the foils exhibited slower reaction propagation. At higher potentials, the propagation speed increased. Too low an ignition potential led to poor propagation, while too high a potential increased the tendency to unstable reaction propagation. Two different types of reaction mode, steady-

Materials Research Forum LLC
doi: http://dx.doi.org/10.21741/9781644900093

state and unsteady, were exhibited by a 1Ti/3Al foil. In the case of unsteady reaction, the reaction temperature and velocity decreased and a macroscopic ripple band structure was observed. Such bands contained a distribution of coarse and fine protrusion microstructures. Melting of the aluminium in the multilayer was expected to affect the unsteady reaction mode in Al-rich multilayer foil. The 1Ti/1Al and 1Ti/2Al foils underwent steady-state reaction with no formation of ripple bands. The time-dependent temperature profile (figure 52) controlled the reaction kinetics, in that thermal transport was faster than atomic mixing.

Table 13. Reaction propagation rates of 185-period Ti|Al reactive foils

Structure	Al (at%)	Thickness (nm)	$E_{ignition}$ (V)	V (m/s)
1Ti|1Al	54	10|10	3	0.73
1Ti|1Al	54	10|10	9	2.57
1Ti|1Al	54	10|10	15	1.38
1Ti|1Al	54	10|10	21	1.83
1Ti|2Al	67	10|18	3	1.15
1Ti|2Al	67	10|18	9	1.25
1Ti|2Al	67	10|18	15	1.24
1Ti|2Al	67	10|18	21	1.47
1Ti|3Al	77	10|27	3	0.68
1Ti|3Al	77	10|27	9	0.95
1Ti|3Al	77	10|27	15	1.60
1Ti|3Al	77	10|27	21	1.58

Self-propagating high-temperature synthesis, closely related to bonding, has been used to join TiAl. Titanium, aluminium and carbon powders and an assisting electromagnetic field were used[195]. The finished joint generally comprised three reaction zones: next to the Tail base metal there was a dark Tail reaction-layer on the interface, while TiC particles and Ti-Al compounds were found in the interlayer. Porosity could not be avoided, due to differences in the specific volumes of the products and reactants, to the

Materials Research Forum LLC
doi: http://dx.doi.org/10.21741/9781644900093

evaporation of impurities and to the expansion of any gas which became trapped in the pores of the reactant compact. To avoid this, Ag-Cu brazing foil was placed between the powder compact and the Tail substrate. It was concluded that the presence of molten Ag-Cu in the filler during the reaction improved the wettability of the interlayer against the Tail substrate and completely filled any holes in the reaction products at the interlayer.

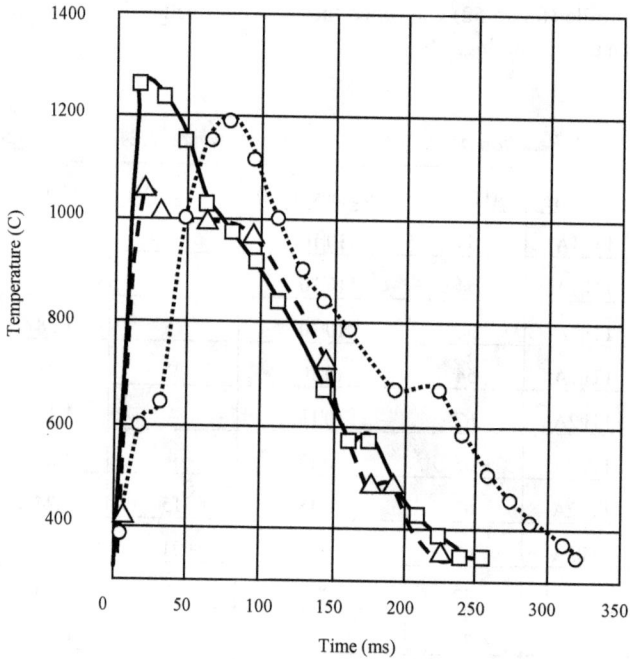

Figure 52. Temperature-time profiles of Ti|Al reactive foils
Circles: 1Ti|1Al, squares: 1Ti|3Al, triangles: 1Ti|2Al

It is again interesting to look at some of the diffusion data which might be relevant to the self-propagating reaction in this system. Attention has been focused on diffusion in β-Ti because of the unusual features of self-diffusion and impurity diffusion of certain elements. These included non-linear Arrhenius plots and very low values of the apparent frequency factor and the apparent activation energy. Measurements of the impurity diffusion coefficient for aluminium in β-Ti had not been carried out because the radioactive tracer method was unavailable; due mainly to the low specific activity of [26]Al,

Materials Research Forum LLC
doi: http://dx.doi.org/10.21741/9781644900093

and to the difficulty of coating this isotope onto a titanium surface. Here, measurements were made[196] of interdiffusion coefficients in the β phase of the Ti-Al system at 1223 to 1573K (table 14), using Ti|Ti-2.1at%Al couples. After annealing, the average grain sizes of the titanium and Ti-2.1at%Al samples were about 2mm, and their oxygen concentrations were less than 600ppm[wt]. The value of the pre-exponential factor (1.2 x $10^{-7}m^2/s$) was lower by a factor of 100 than the value for self-diffusion which had been suggested on the basis of a vacancy mechanism, and indicated a negative activation entropy. On the other hand, the activation energy (150kJ/mol) for the impurity diffusion of aluminium was equal to about 50% of that for self-diffusion which was predicted by an empirical rule. Furthermore, the activation energy was considerably lower than the value for self-diffusion which was predicted by an empirical relationship between the activation energies for diffusion in cubic metals and their latent heats of melting. The very low values of the pre-exponential factor and the activation energy indicated that the impurity diffusion of aluminum in β-Ti could be regarded as being anomalous.

The initial stage of thin-film interdiffusion was studied at 3000K. Examination revealed the presence of Al_3Ni, $Al_{65}Ni_{35}$, $Al_{60}Ni_{40}$, $Al_{56}Ni_{44}$, $Ni_{60}Al_{40}$ and Ni_3Al; even though these intermetallics were not found in the equilibrium phase diagram. [197]

Table 14. Diffusivity of aluminium in β-Ti

Temperature (K)	D (m^2/s)
1573	1.2×10^{-12}
1523	9.4×10^{-13}
1473	6.4×10^{-13}
1423	3.3×10^{-13}
1373	2.1×10^{-13}
1323	1.3×10^{-13}
1273	8.6×10^{-14}
1223	4.9×10^{-14}

Self-diffusion and aluminium impurity diffusion were studied[198] in the hexagonal close-packed α-phase. Samples were used which had various impurity contents. These included

Materials Research Forum LLC
doi: http://dx.doi.org/10.21741/9781644900093

ultra-pure material with extremely low concentrations of interstitial impurities such as iron, cobalt and nickel. Self-diffusion measurements were performed by using a [44]Ti radiotracer and the ion-beam sputtering technique. In-depth profiling by secondary ion mass spectrometry was used for the aluminium diffusion measurements. These were made both perpendicular to, and parallel to, the c-axis; using single crystals and coarse-grained polycrystals (table 14). The results for the ultra-pure α-phase, perpendicular to the c-axis, could be described by:

$$D \ (m^2/s) = 6.6 \times 10^{-3} \exp[-329(kJ/mol)/RT]$$

The ratio of the parallel diffusivities to the perpendicular diffusivities was equal to about 0.5 for self-diffusion and equal to about 0.65 for aluminium diffusion. These results were treated as being the intrinsic diffusion properties of α-Ti. They were consistent with the normal diffusion behavior in other hexagonal close-packed metals. It was concluded that both self-diffusion and substitutional solute diffusion in α-Ti were intrinsically normal and were dominated by the vacancy mechanism. Diffusion in less pure material was more rapid and required a lower activation energy. This was attributed to an enhancement of atomic mobility in the matrix, due to interstitially dissolved fast-diffusing impurities.

Table 15. Diffusivity of aluminium in α-Ti single crystals

Temperature (K)	D (m^2/s)	Orientation*
935	2.17×10^{-21}	$\perp$
973	1.28×10^{-20}	$\perp$
1010	5.92×10^{-20}	$\perp$
1036	2.14×10^{-19}	$\parallel$
1036	3.95×10^{-19}	$\perp$
1050	2.44×10^{-19}	$\perp$
1073	7.67×10^{-19}	$\perp$
1073	5.02×10^{-19}	$\parallel$
1093	1.15×10^{-18}	$\perp$
1140	5.42×10^{-18}	$\perp$

* wrt c-axis

The diffusion of implanted solute in the α-phase was studied[199] at 948 to 1073K by using nuclear reaction analysis. The measurements showed that the diffusion coefficients yielded a linear Arrhenius plot. The results could be described by:

$$D \ (m^2/s) = 1.4 \times 10^{-2} \ exp[-326(kJ/mol)/RT]$$

These parameters were typical of normal substitutional behavior. Reaction diffusion in Ti|Al and Ti-5mol%O|Al diffusion couples was investigated[200] by means of electron-probe micro-analysis. Only Al_3Ti was observed, as an intermediate layer, in both couples at 773 to 903K. The silicon, an impurity element in aluminium, was concentrated in the Al_3Ti. The growth of the intermediate layer was diffusion-limited, with an estimated activation energy of 237kJ/mol. In the case of O-doped diffusion couples, the layer growth of Al_3Ti was appreciably suppressed and the activation energy was equal to 263kJ/mol at 773 to 873K. This suppression was attributed to the presence of Al_2O_3 which formed between the Al and the Al_3Ti. Kirkendall markers shifted towards the aluminium side; thus suggesting that the diffusion of aluminium was faster than that of titanium in the intermediate layer.

The diffusion of aluminum in polycrystalline samples of the α phase was measured[201] by using nuclear resonance broadening methods at 600 to 850C. The results could be described by:

$$D \ (cm^2/s) = 7.4 \times 10^{-7} \ exp[-1.62(eV)/kT]$$

Solid-state reactive diffusion between titanium and aluminium was investigated[202] at 520 to 650C by using multi-laminated Ti|Al diffusion couples. This sort of study is obviously much closer to the bonding situation. In samples which were annealed at up to 150h, $TiAl_3$ was the only phase which was observed in the diffusion zone. The preferential formation of this compound in Ti|Al diffusion couples was predicted by using an effective heat-of-formation model. The present results indicated that titanium and aluminium diffused into each other, and that the growth of $TiAl_3$ layers occurred mainly towards the aluminium side. The $TiAl_3$ growth kinetics changed from parabolic to linear between 575 and 600C, and were characterized by activation energies of 33.2 and 295.8kJ/mol, respectively. It was suggested that the low-temperature kinetics were dominated by the diffusion of titanium atoms along the grain boundaries of the $TiAl_3$ layers. Reaction at $TiAl_3$|Al interfaces in the high-temperature regime was limited by the diffusion of titanium atoms in the aluminium foils. This was the result of an increasing solubility of titanium in aluminium with increasing temperature.

On titanium surfaces coated by an aluminium film, various aluminide phases formed during annealing as the result of reaction between the titanium and aluminium[203].

Materials Research Forum LLC
doi: http://dx.doi.org/10.21741/9781644900093

Preliminary irradiation, of aluminium film with a thickness of 7μm, by Ti^+ ions had a strong effect upon the interdiffusion growth of aluminide phases on the titanium substrate. Preliminary ion irradiation resulted in the development of more homogeneous and fine-grained microstructures during subsequent annealing. During ion irradiation of the two-phase ($TiAl+Ti_3Al$) overlayer, decomposition of the TiAl compound and the formation of Ti_3Al occurred. During subsequent annealing, diffusion cementation of the overlayer occurred faster on the surface of irradiated samples. Following bombardment with various ions (Ti^+ and Al^+), and during subsequent annealing, the kinetics of structural formation developed differently.

Electrometric measurements[204] of the oxidation of Ti|Al thin-film couples revealed that the titanium diffusivity at 400C was between 1×10^{-15} and $4 \times 10^{-15} cm^2/s$. At 450C, it was between 4×10^{-15} and $1.6 \times 10^{-14} cm^2/s$. This implied that the diffusivity between 400 and 450C could be described by:

$$D(cm^2/s) = 5 \times 10^{-7} exp[-1.12(eV)/kT]$$

Other electrometric measurements[205] of the diffusion of titanium into 225 to 250Å thin films revealed that the titanium diffusivity at 690K was $2.9 \times 10^{-18} cm^2/s$. and that the diffusivity between 650 and 740C could be described by:

$$D(cm^2/s) = 1.9 \times 10^{-5} exp[-1.74(eV)/kT]$$

The interaction of titanium and aluminium layers was monitored[206] by resistometric measurements, and was correlated with the creation of intermetallic compounds. The diffusion activation energy increased as the reaction rate of the intermetallics decreased. The diffusivity of aluminium in titanium decreased from 6×10^{-14} to $6.6 \times 10^{-17} cm^2/s$. (figure 54).

The diffusivity of aluminium in alpha-titanium (table 15) was determined[207] by vapor-depositing aluminium layers, 0.1 to 0.4μm thick, onto 0.8mm-thick titanium foil, and could be described by:

$$D(cm^2/s) = 9.7 \times 10^{-5} exp[-27.5(kcal/mol)/RT]$$

between 700 and 850C.

Materials Research Forum LLC
doi: http://dx.doi.org/10.21741/9781644900093

Table 16. Diffusivity of aluminium in alpha-titanium

Temperature (C)	Diffusivity (cm^2/s)
700	6.6 x 10^{-11}
750	1.4 x 10^{-10}
800	2.5 x 10^{-10}
850	4.3 x 10^{-10}

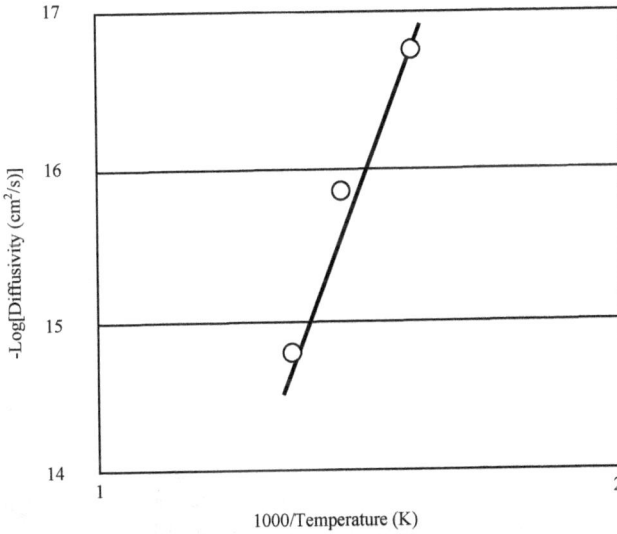

Figure 54. Diffusivity of aluminium in aluminium\titanium layers

The reaction of 0.0005 to 0.001mm aluminium films with thick substrates to form TiAl₃ was such that it occurred rapidly at 635C and was confined to the surface region[208]. When heated at 900C, Ti₃Al formed with little release of aluminium into the alpha-titanium. Continued annealing at that temperature then caused the Ti₃Al to decompose, releasing a large amount of aluminium into the titanium. The interdiffusion coefficient of aluminium

Bonding by Self-Propagating Reaction
Materials Research Foundations 45 (2019)

Materials Research Forum LLC
doi: http://dx.doi.org/10.21741/9781644900093

in titanium at 900C increased by less than an order of magnitude as the aluminium content varied from 0 to 20at% (figure 55).

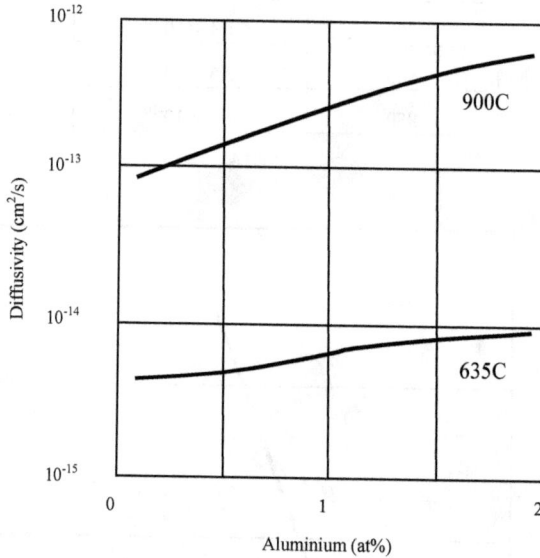

Figure 55. Interdiffusion of aluminium and titanium as a function of composition

A study of the behaviour of Ti|Al-Cu|Si films showed[209] that the addition of silicon and/or copper to aluminium led to a retarded reaction between aluminium and titanium during annealing, with titanium diffusion occurring throughout the entire aluminium film. In the absence of silicon or copper, the titanium content was homogeneous. The addition of silicon did not prevent titanium diffusion, but reduced the titanium concentration in the film bulk while producing a higher concentration in the near-surface region. The addition of copper produced a much less marked effect.

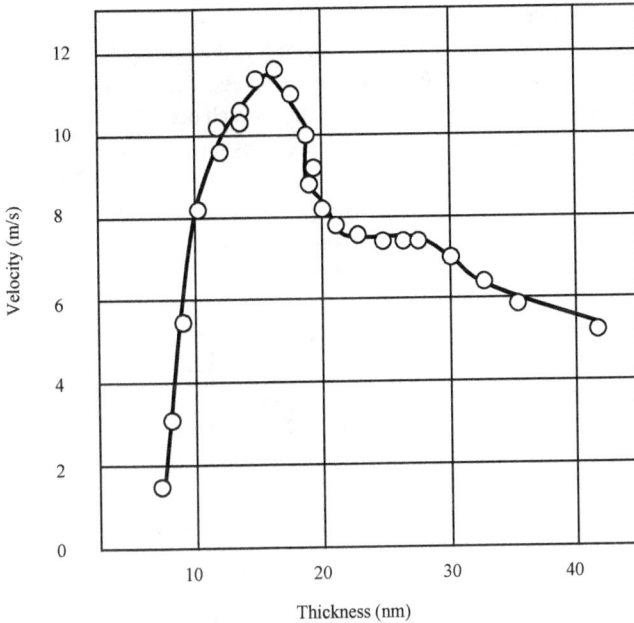

*Figure 56. Propagation velocity as a function of the
bilayer thickness of multilayered 2Al\1Zr foils*

Al\Zr

High-temperature self-propagating reactions were observed in vapor-deposited multilayered nanoscale foils having overall atomic ratios of 3:1 or 2:1. The reaction velocities (figure 56) did not exhibit the expected inverse dependence upon bilayer thickness that was predicted on the basis of changes in the average diffusion distance. For bilayer thicknesses of 20 to 30nm, the velocity was constant and equal to about 7.7m/s. The reaction temperature attained a constant value (figure 57) and roughly paralleled the heat-of-reaction (figure 58). The phase evolution during self-propagating reaction in foils with 3:1 stoichiometry involved rapid transformation from Al\Zr multilayer to equilibrium Al_3Zr, with no intermediate crystalline phases, and probably involved a molten aluminium-rich phase. The phase evolution was the same for 90nm-thick bilayer

Materials Research Forum LLC
doi: http://dx.doi.org/10.21741/9781644900093

foils and for bilayer foils having thicknesses ranging from 27 to 35nm. For foils with a bilayer thickness of 90nm and a 3:1 overall chemistry, the propagation front was planar and steady[210]. In foils having a 90nm bilayer thickness, the reaction propagated as a steady front; unlike the oscillatory unsteady reaction which occurs in 1Al|1Zr foils having a similar bilayer thickness. Among the possible reasons for the above plateau in velocity values, were a transition from steady to unsteady propagation mode, a change in the sequence of phase transformations and an hexagonal close-packed to body-centred cubic transformation in the zirconium reactant layer. None of these explanations were fully satisfactory.

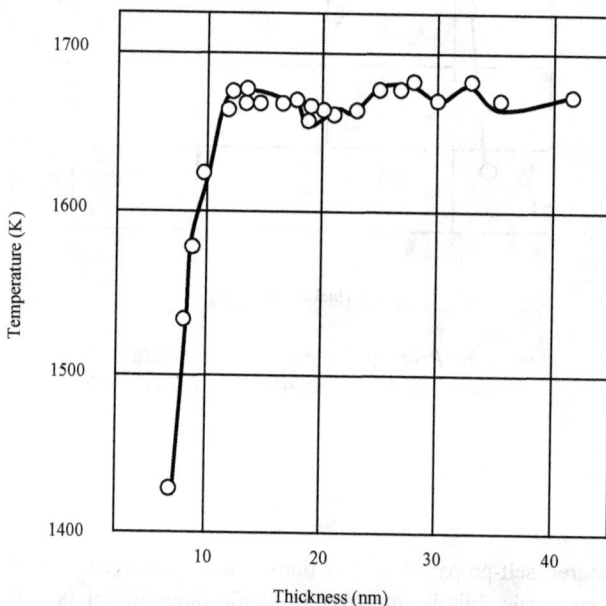

Figure 57. Reaction temperature as a function of the bilayer thickness of multilayered 2Al|1Zr foils

Unsteady reaction fronts occurred in foils with a 1:1 overall chemistry and similar bilayer thickness. Magnetron sputter-deposited multilayers of the form, nAl|Ti (n = 1 to 3), having various aluminium molar ratios and of the form, 1Al|1Zr, having bilayer thicknesses of 20 to 55nm were deposited onto a copper substrate. In order to fabricate

Materials Research Forum LLC

doi: http://dx.doi.org/10.21741/9781644900093

thick films, two different processes were used: continuous coating and coating with interval etching. The materials had a columnar microstructure, with coherent reactant layers. The interval etching process led to consistent and uniform film growth throughout the cross-section, as compared with the other process. Free-standing specimens of Al|Zr which were about 10cm long and 45µm thick, with homogeneous film growth, could be prepared.

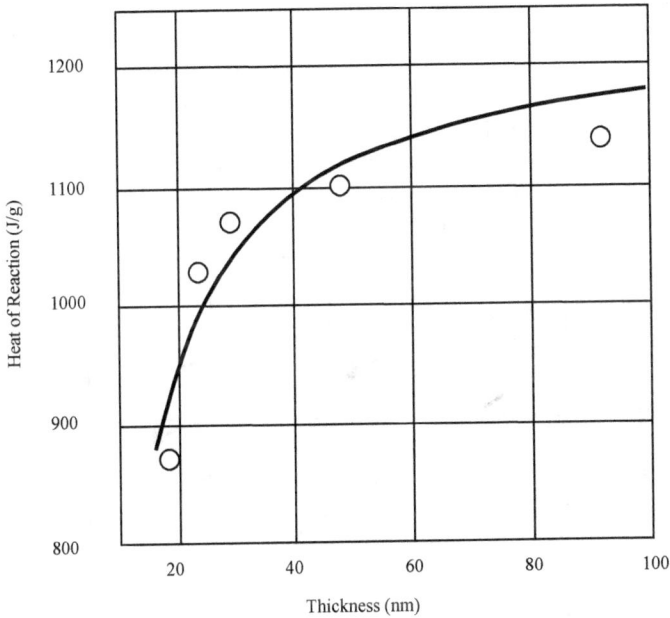

*Figure 58. Heat of reaction as a function of
the bilayer thickness of 3Al|1Zr foil*

Both Al|Ti and Al|Zr multilayer foils exhibited a maximum reaction velocity of 2.6 and 1.22m/s, respectively[211]. Maximum reaction temperatures of 1581 to 1707C were found for Al|Zr, while the nAl|Ti multilayer foils had reaction temperatures of 1215 to 1298C. The foils exhibited self-propagating reactions following 9V electrical ignition in air. The propagation rate depended upon the reactant-pair involved. For a given system, the reaction propagation and temperature were greatly influenced by the bilayer thickness and molar ratio. Both Zr|Al and Ti|Al films had relatively low reaction velocities as compared with that of Ni/Al nanofoil, but the reaction temperature of the Zr|Al was higher than that of the nanofoil.

A computational model has been developed, in order to describe the oxidation of nanolaminates which include Al|Zr bilayers, prompted by experimental observations of reactive multilayer ignition in air. The model predictions suggested that, in the early stages following completion of the reaction, oxidation was best described as being surface-reaction controlled growth[212]. A transition to diffusion-controlled growth occurred as the oxide layer thickened. A further simplified model was able to incorporate both oxide-growth regimes. Evolution of the foil temperature was described by an energy-balance equation which took account of the heat of oxidation, oxygen intake and radiative heat loss. The results could be used to estimate the oxidation-heat release rate, the temperature of the oxidizing foil and the effect of radiative heat losses.

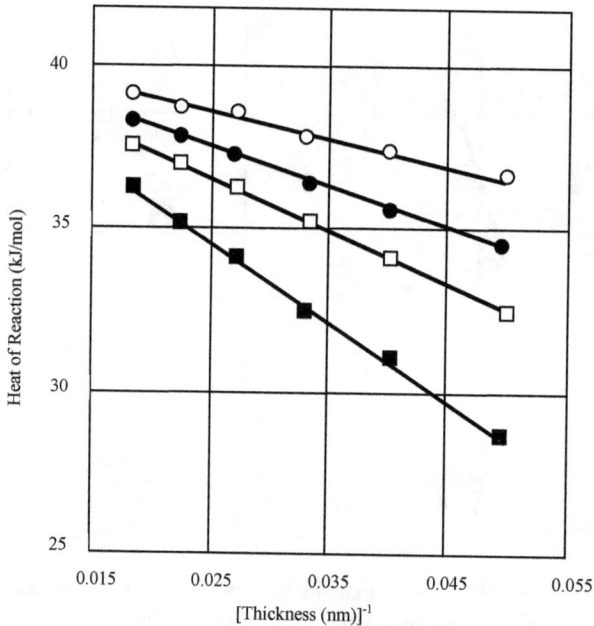

Figure 59. Reaction heat of 1Zr|1Al reactive films as a function of inverse bilayer thickness and pre-mixed layer thickness Open circles: 1nm, solid circles: 1.5nm, open squares: 2nm, solid squares: 3nm

Bonding by Self-Propagating Reaction
Materials Research Foundations **45** (2019)

Materials Research Forum LLC
doi: http://dx.doi.org/10.21741/9781644900093

In situ X-ray diffraction of a 40µm-thick Al|Zr multilayer foil was used to monitor the growth of reaction products during the ignition, reaction and cooling (during about 5s) in order to determine the mechanisms which prevent complete reaction from occurring[213]. The simultaneous use of pyrometry (figure 59) made it possible to relate the phase formation to the foil temperature during reaction (figure 60). It was found that the main reaction product, orthorhombic ZrO_2, grew linearly during the first 1.3s of reaction; suggesting the occurrence of interface-limited growth. It then changed to slower diffusion-controlled growth. This transition in growth rate (figure 61) was associated with the sudden end of a temperature plateau which was associated with self-sustained reaction. A 8µm-thick Al_2Zr layer was found beneath the oxidized exterior and was blamed for the suppression of the reaction before complete reaction had occurred. The latter Al-rich layer prevented the diffusion of aluminium away from the oxide interface; thus causing an increasing fraction of the aluminium to oxidize. The resultant alumina was then a barrier to oxygen diffusion and led to incomplete reaction.

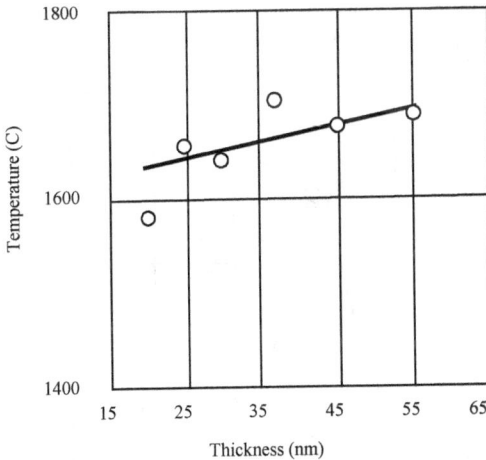

Figure 60. Maximum reaction temperatures of Zr|Al multifoils in air

It was noted that Al|Zr and Al-8Mg|Zr nanocomposite foils did not react completely in air because the degree of penetration of oxygen and nitrogen into the foils tended to become limited as the product phases formed. It was suggested that the heat which was produced during foil reaction could depend upon the volume fraction of the surface oxide that formed and therefore upon the initial foil thickness. Some foils of Al|Zr and Al-8Mg|Zr, with thicknesses of 9 to 61μm, were prepared using physical vapor deposition. The reaction efficiency decreased appreciably for Al|Zr foils, as the thickness increased, but the Al-8Mg|Zr foils exhibited an essentially constant reaction efficiency.

In Al|Zr foils, a distinct oxide layer formed on the external surfaces and there was a low level of oxygen and nitrogen towards the center. In Al-8Mg|Zr foils there was hardly any correlation between the heat output and the foil thickness. The surface oxide layer was here more diffuse, and the oxygen and nitrogen contents were higher throughout the foil. It was concluded that the addition of magnesium improved heat generation by increasing the rates of oxygen and nitrogen diffusion[214]. This permitted the formation of solid solutions that were richer in oxygen and nitrogen throughout the bulk of the foil.

The effect of magnesium upon the ability of nanocomposite foils to generate heat was determined by comparing Al|Zr (figure 62) and Al-8Mg|Zr (figure 63). Measurements showed that the combustion temperature and duration were such that, for both foil compositions, the temperature was not greatly affected by the foil geometry although the burn duration increased slightly with increasing thickness for both foils.

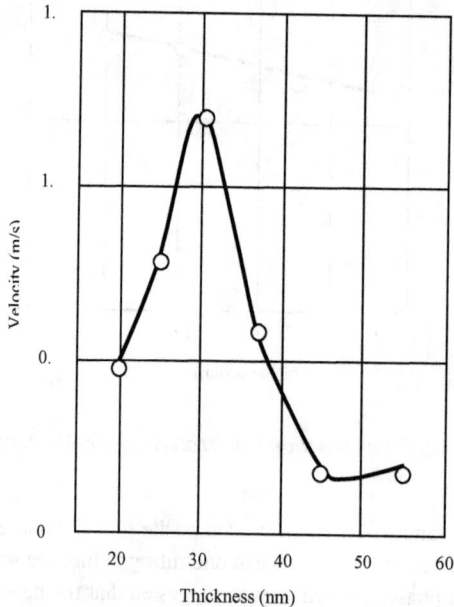

Figure 61. Reaction front propagation velocity in 1Zr|1Al multifoil

It was suggested that these effects could be explained by assuming that the magnesium additions in Al-8Mg|Zr produced vacancies which increased the diffusion rates of gaseous reactants into the foil. This then permitted the maintenance of a greater combustion efficiency in thicker foils. The oxygen and nitrogen then formed solid solutions in the central intermetallic regions of the Al-8Mg|Zr foils, thus releasing sufficient heat to offset the decreasing contribution arsing from the external oxide layers as the foil thickness increased. There was thus a strong dependence upon the geometry of Al|Zr foils because the volume fraction of the oxide shell became ever smaller as the foil thickness increased. The equivalent geometrical dependence was insignificant for Al-8Mg|Zr foils, when they were more than 10µm thick, because the oxygen and nitrogen could penetrate the foil thickness more easily.

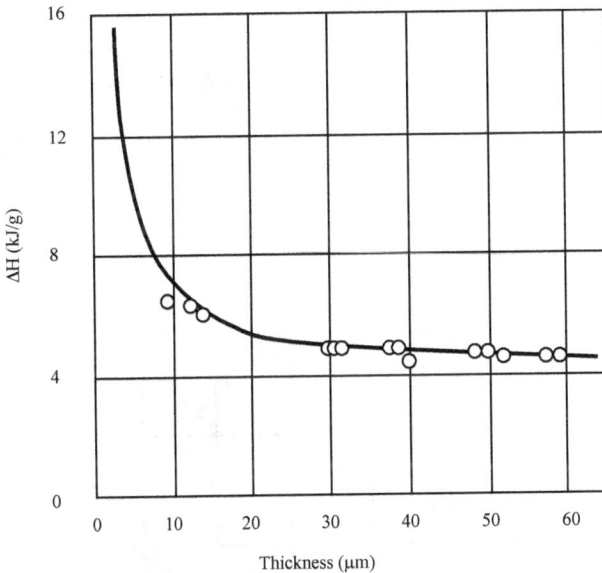

Figure 62. Heat output of Al\Zr foil as a function of foil thickness

Measurements performed using a bomb calorimeter showed that the Al-8Mg|Zr foils generated the most gravimetric heat in air, oxygen or nitrogen ambients[215]. Both of the magnesium-containing foils released a plume of particles and vapor upon reacting. The ejected mass included magnesium vapor, and particles of all of the metals. Both the vapor

Materials Research Forum LLC
doi: http://dx.doi.org/10.21741/9781644900093

and the particles oxidized rapidly in air, resulting in single metal-oxide particles. The reacted foils, especially in the case of the 8at%Mg samples, contained voids and higher levels of oxygen and nitrogen. This higher heat-of-reaction suggested, as above, that the out-diffusion and evaporation of magnesium generated a high concentration of vacancies which increased oxygen and nitrogen diffusion within the foil; thus increasing the extent of oxidation and nitridation.

The high-temperature oxidation behavior during fast self-propagating reaction was further studied in binary Al|Zr multilayer films. Magnetron-sputtered free-standing films having various bilayer thicknesses were investigated with regard to oxidation behavior[216]. This revealed the formation of oxides of zirconium and aluminium, together with intermetallic compounds. Oxidation was reduced by an argon atmosphere, but a minute amount of ZrO_2 alone formed. The temperature profiles exhibited a temperature plateau after reaching the maximum temperature. The static plateau temperatures ranged from 1475 to 1588C, depending upon the bilayer thickness; the temperature plateau becoming more marked with increasing bilayer thickness. A longer period spent at the high-temperature plateau was characteristic of appreciable oxidation in air during self-propagating reaction.

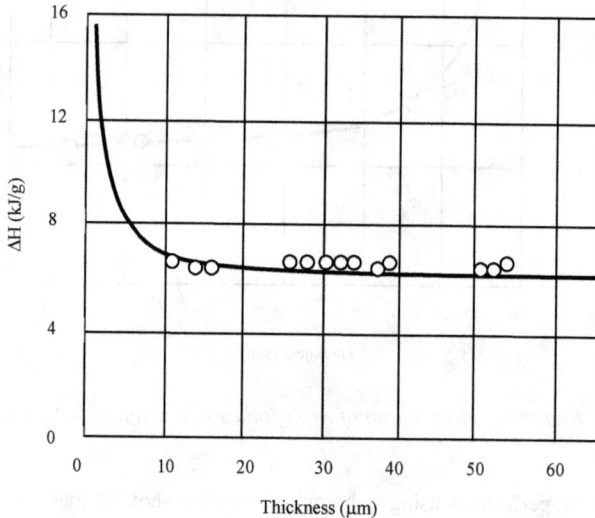

Figure 63. Heat output of Al-8Mg|Zr foil as a function of foil thickness

Materials Research Forum LLC
doi: http://dx.doi.org/10.21741/9781644900093

The energies and sequences of phase transformations which occurred in multilayered foils with the atomic ratio, 3Al:1Zr, were studied during heat-treatment at up to 350C in a differential scanning calorimeter[217]. The first phase to be formed was an Al-rich amorphous phase which apparently grew via zirconium diffusion through the amorphous phase. Subsequent nucleation and growth of tetragonal Al_3Zr along the aluminium|amorphous-layer interface was controlled by aluminium diffusion through the crystalline intermetallic phase. The diffusion coefficients which were associated with these processes were higher than were to be expected on the basis of reported diffusivities for higher temperatures. The heat-of-formation of the tetragonal Al_3Zr phase was deduced to be 1240J/g. No anomalous variation in the energies or sequences of the phase transformations was observed for bilayer thicknesses of between 17 and 90nm, even though there were anomalies in the bilayer-dependence of self-propagating reaction velocities in the same foils.

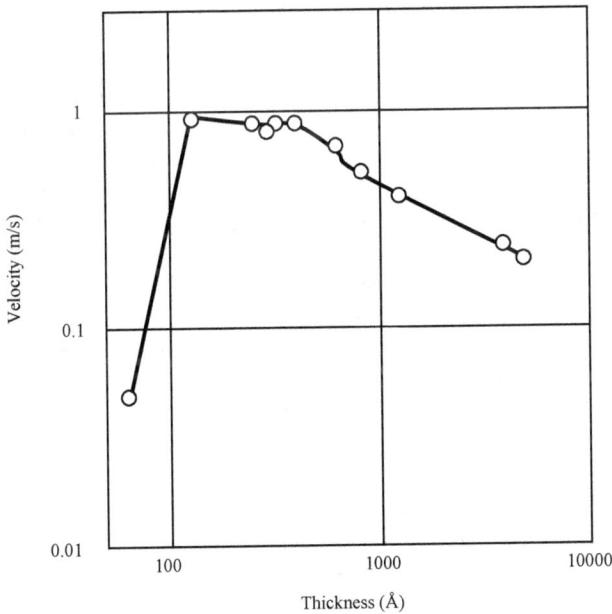

Figure 64. Propagation speed as a function of
5 μm Ni\Ti nanolaminate bilayer thickness

Materials Research Forum LLC
doi: http://dx.doi.org/10.21741/9781644900093

Again looking at the solid-state diffusion aspects, the diffusion of aluminium and zirconium was investigated[218] in the β-Zr(Al) phase at 1203 to 1323K by using single-phase diffusion couples of pure zirconium against Zr–2.8wt%Al. The interdiffusion coefficients exhibited a small increase with increasing aluminium concentration, and obeyed a quadratic compositional relationship. The temperature dependence of the interdiffusion coefficients at various compositions was established. The activation energy for interdiffusion decreased linearly with increasing aluminium concentration. The intrinsic diffusivity of zirconium was higher than that of aluminium in this phase field. The impurity diffusion coefficient of aluminium in β-Zr was determined by extrapolating interdiffusion coefficients to a limiting concentration of aluminium, and was found to exhibit a temperature dependence of the form:

$$D \ (m^2/s) = 5.567 \ x \ 10^{-6} exp[-220.08(kJ)/RT]$$

A correlation between the impurity diffusion coefficients of various impurities in β-Zr, and the atomic radii of the impurity atoms, was established and could be described by the relationship:

$$\log[D \ (m^2/s)] = -14.57 + exp[4.84 - 30.22r(nm)]$$

Solid-state diffusion reaction was studied[219] at 838 to 898K by using bulk diffusion couples. It was found that $ZrAl_3$ formed in the diffusion zone in all of the couples, with a planar morphology. A second layer of Zr_2Al_3 also formed in the diffusion zone, at and above 873K. The formation of $ZrAl_3$ as the first phase was explained on the basis of the modified effective heat of formation model. The driving forces for the formation of Zr_2Al_3 and $ZrAl_2$ at the $Zr|ZrAl_3$ interface were evaluated. The appearance of Zr_2Al_3, in preference to $ZrAl_2$, at higher temperatures was attributed to the higher driving force for its formation at the $Zr|ZrAl_3$ interface. The temperature dependence of the growth of the $ZrAl_3$ phase layer was determined. The activation energy for the growth of $ZrAl_3$ was 188.2kJ/mol, and the pre-exponential factor was $2.9 \ x \ 10^4 m/s^{1/2}$.

Ni|Ti

Nanolaminates which were fabricated by sputter-deposition exhibited rapid high-temperature reaction. When locally heated, self-sustained reactions occurred in free-standing Ni|Ti multilayer foils and propagated (figure 64) at average speeds of between 0.1 and 1.4m/s[220].

This speed was affected by the total foil thickness and layer periodicity. No pre-heating was required in order to maintain the reaction. Stimulation by rapid global heating indicated ignition temperatures (figure 65) of 300 to 400C for nanolaminates, and the ignition temperature was affected by the bilayer thickness. Coarser laminates required a higher ignition temperature. Foils which were reacted under vacuum generated either single-phase B2 cubic NiTi or a mixture of monoclinic B19′ NiTi, hexagonal $NiTi_2$ and B2 NiTi. Multilayers with a thickness greater than 28nm form this mixture of phases. The relative amount of austenite decreases with increasing bilayer thickness, as the concentrations of $NiTi_2$ and B19′ martensite increase. The incomplete blending of reactants in thicker bilayers was attributed to limited reaction kinetics. Micrographs reveal discontinuous layers of Ti-rich precipitates, with an out-of-plane periodicity that is about the same as the original bilayer thickness, within a matrix of NiTi.

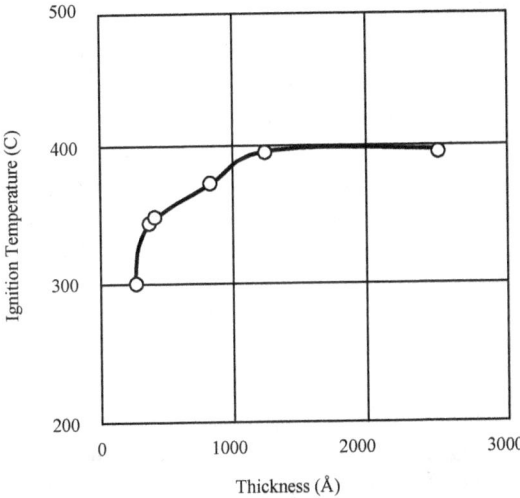

Figure 65. Ignition temperature of Ni\Ti nanolaminate
as a function of bilayer thickness

A ten-fold increase in the laser spot diameter could reduce the ignition threshold fluence by some two orders of magnitude. Nickel|titanium nanolaminates which were made by sputter-deposition underwent rapid high-temperature synthesis. Upon heating locally, self-sustaining reaction in free-standing nickel/titanium multilayer foils led to average propagation speeds of 0.1 to 1.4m/s. As usual, the speed of propagation was affected by the total foil thickness and the bilayer thickness. Unlike other work on compacted nickel-titanium powder, no pre-heating of the foils was required in order to maintain self-propagating reaction. High-temperature synthesis was also stimulated by rapid global heating, leading to ignition temperatures of only 300 to 400C for nanolaminates. Coarser

Materials Research Forum LLC
doi: http://dx.doi.org/10.21741/9781644900093

laminates required a higher ignition temperature. Foils which were reacted *in vacuo* developed a single-phase B2 cubic NiTi structure or a mixed-phase structure which comprised monoclinic B19′ NiTi, hexagonal NiTi$_2$ and B2 NiTi. Monophase cubic B2 NiTi tended to form when the initial bilayer thickness was small. Vapor-deposited nickel|titanium multilayer foils with a net 1:1 stoichiometry exhibited self-propagating high-temperature combustion with steady-state propagation rates, for free-standing foils, ranging from 0.2 to 1.0m/s. The films had a fine-grained polycrystalline microstructure and, as found elsewhere, comprised cubic B2 and monoclinic B19′ phases plus some NiTi$_2$ or Ni$_3$Ti precipitates[221].

Single-phase cubic B2 NiTi usually formed when the initial bilayer thickness was minimized. In order to boost the ability of this metallic combination to melt its surroundings, even if their melting point exceeds 500C, a useful modification is to supplement the internal energy of the Ni|Ti layer by using externally applied microwave energy. This technique has been used to bond diamond to tungsten carbide[222]. The microwave energy was used to heat the whole assembly, whereupon the reactive multilayer ignited at the center of the sandwich arrangement. A drawback of this combination is that the reaction can be stifled by conductive adjacent material and may react successively only if detached from neighboring substances. The 9at%V which is often added to nickel[223,224] reacts to form the minor phase, Al$_6$V.

The bonding of TiAl alloy to AISI310 stainless steel has been carried out by using Ni|Ti reactive multilayers which were alternately deposited, using direct-current magnetron sputtering, onto the materials to be joined. The bilayer thickness was 30 or 60nm and bonding was performed at 700 or 800C for 1h under pressures of 50 or 10MPa. The joint interfaces were approximately 5μm thick, and had a layered microstructure. They were composed mainly of equiaxed grains of NiTi and NiTi$_2$. The thickness and the number of layers depended upon the bonding conditions and the bilayer thickness of the multilayers[225]. The Young's modulus distribution highlighted the phase differences which existed across the joint interface. The highest joint shear strength resulted from bonding at 800C for 1h under a pressure of 10MPa, while using Ni|Ti multilayers having a 30nm bilayer thickness.

Bonding of TiNi to Ti-6Al-4V was carried by using reactive Ni|Ti multilayer thin films. The TiNi and Ti-6Al-4V surfaces were modified, as above, by sputter-depositing alternating nickel and titanium nanolayers in order to increase the diffusivity at the interface. Bonding was carried out at 750, 800 and 900C using a pressure of 10MPa and a dwell-time of 1h. Joints which were free from porosity and cracking were produced[226]. The reaction zone contained columnar grains of Ti$_2$Ni and AlNi$_2$Ti, close to the Ti-6Al-

Materials Research Forum LLC

doi: http://dx.doi.org/10.21741/9781644900093

4V, and alternating layers of Ti_2Ni and TiNi equiaxed grains. The grain size decreased in going from the Ti-6Al-4V to the TiNi. Nanometric grains occurred in the layers which were closest to the TiNi.

X-ray diffraction using synchrotron radiation has been used for the real-time investigation of phase evolution in Ni|Ti multilayer thin films during annealing[227]. The multilayers were direct-current magnetron-sputtered onto Ti-6Al-4V substrates using pure nickel and titanium targets. The deposition parameters were such as to yield a near-equiatomic chemical composition and modulation periods shorter than 25nm. Well-defined structures with alternating nickel-rich and titanium-rich layers were observed along the whole thickness of the films, even for a modulation period of 4nm. The halo which characterized an amorphous structure was observed. For a modulation period of 12 or 25nm, the as-deposited thin films were nanocrystalline and exhibited (111) nickel and (002) titanium diffraction peaks. The nanolayer structure disappeared during annealing, due to interdiffusion followed by reaction. The reaction of nickel and titanium to produce cubic B2-structured NiTi occurred within a short time and within a narrow temperature range. For a modulation period of 25, 12 or 4nm, the reaction temperature was close to 320, 350 or 385C, respectively. At higher temperatures, $NiTi_2$ was also found and titanium diffusion from the substrate plus nickel towards the substrate was expected to favour the precipitation of $NiTi_2$.

Again looking at possibly relevant diffusion data, samples were prepared by the alternate evaporation of nickel and titanium on (111) silicon substrates. Interlayer diffusion and solid-state reaction in the multi-layer films were then studied, at between 523 and 623K, by means of Auger electron spectroscopic depth profiling and grazing-incidence X-ray diffractometry. The resultant data for nickel diffusion could be described[228] by:

$$D \ (m^2/s) = 3 \times 10^{-11} \ exp[-120(kJ/mol)/RT]$$

and that for titanium diffusion by:

$$D \ (m^2/s) = 7 \times 10^{-10} \ exp[-130(kJ/mol)/RT]$$

Diffusion in Ni-Ti multi-layers with periods of 12nm was studied at 293 to 743K by means of grazing-angle non-polarized neutron reflectometry. The effective diffusion coefficients of nickel into titanium (table 17), and the corresponding activation energy, were deduced from measurements of the decay of the reflectivity of the first Bragg peak, which arose from nuclear scattering-length density modulations, as a function of isochronal annealing. The diffusion direction was determined[229] by simulating the Kiessig fringes which were located between the total-reflection plateau and the first Bragg peak. Two diffusion regimes, with a transition temperature of about 543K, were observed in this Ni|Ti multi-layer. The corresponding activation energies were 0.21 and 0.43eV, respectively.

Table 17. Diffusivity of Ni in Ti as a function of temperature

Temperature (K)	D (cm^2/s)
389	1.55×10^{-17}
443	2.46×10^{-17}
493	3.53×10^{-17}
543	1.543×10^{-16}
593	2.007×10^{-16}
643	3.048×10^{-16}
693	3.463×10^{-16}
743	3.693×10^{-16}

Table 18. Diffusivity of nickel in titanium within a magnetic field

Temperature (K)	Strength (T)	Orientation*	Diffusivity (m^2/s)
1112.1	4	parallel	1.198×10^{-12}
1112.1	4	perpendicular	1.207×10^{-12}
1115.2	4	perpendicular	1.183×10^{-12}
1115.6	0	-	1.143×10^{-12}
1179.9	4	parallel	4.102×10^{-12}
1179.9	4	perpendicular	4.128×10^{-12}
1179.3	0	-	3.905×10^{-12}

*with respect to the field direction

The laws of melt formation, and subsequent solid-liquid interaction between nickel substrates and titanium thin films at 1050, 1150 and 1250C (for 0.25 to 1h), were studied[230]. This permitted solid-phase diffusion of titanium from the melt, and into the nickel substrate, to be investigated. At 1250C, for a solid solution of titanium in nickel

Materials Research Forum LLC
doi: http://dx.doi.org/10.21741/9781644900093

and parabolic layer growth, the growth-rate constant was 2.1 x $10^{-12}m^2$/s and the mutual diffusion coefficient was equal to 1.8 x 10^{-13} m^2/s. Transmission electron microscopy of the solid-state reaction of multilayered thin films showed that the reaction product was a simple intermetallic compound[231]. Autoradiographic analysis showed[232] that the diffusion of ^{63}Ni in technical-grade titanium between 950 and 1050C could be described by:

$$D(cm^2/s) = 1.8 \times 10^5 exp[-76.3(kcal/g\text{-}atom)/RT]$$

It has been de000monstrated[233] that a magnetic field has little effect upon the diffusivity (table 18).

Interaction of face-centered cubic and hexagonal close-packed multilayers, with modulation wavelengths of 3, 5, 8 or 11nm, at 453 to 513K indicated[234] diffusion coefficients ranging from 6 x 10^{-25} to 7 x $10^{-24}m^2$/s. (figure 66).

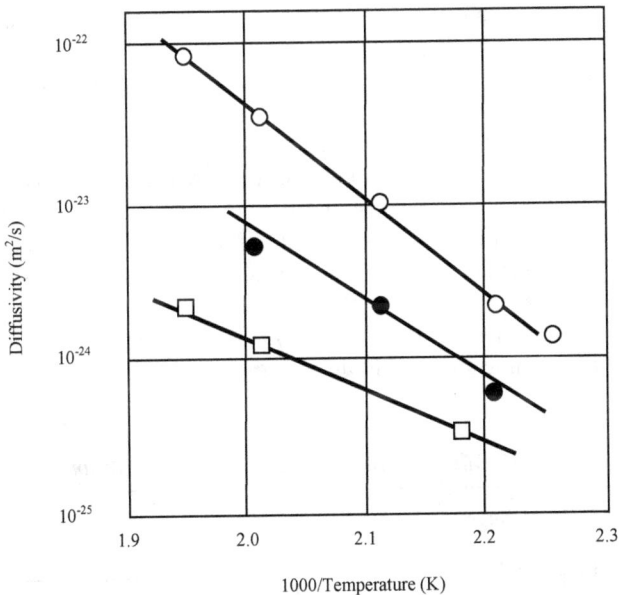

Figure 66. Interdiffusion in nickel\titanium multilayers with modulation wavelengths of open circles, 11nm, solid circles, 8nm, and squares, 3.5nm

Ni|Zr

Intermetallic formation was studied, in 70nm vapor-deposited multilayered foils with overall atomic ratios of 1:1, 2:1 or 7:2, using a heating rate of 1K/s. All three stoichiometries first formed a Ni-Zr amorphous phase which initially crystallized to give NiZr. The heat of reaction up to the final phase was 34 to 36kJ/mol in every case. Intermetallic formation was also studied while using a heating rate of more than 10^5K during high-temperature self-propagating reaction. The reaction velocity and maximum reaction temperature were essentially independent of the foil stoichiometry and were 0.6m/s and 1220K, respectively[235]. The maximum reaction temperature (table 19) was over 200K lower than the predicted adiabatic temperature, and this discrepancy was attributed to the fact that transformation to the final intermetallic phase occurred after the maximum reaction temperature and resulted in the release of 20 to 30% of the total heat of reaction and hence to a delay in cooling. Multiple phases are present following reaction because more than one phase is stable for the chosen composition. Multilayers with the composition, 2Ni|Zr form a mixture of $Ni_{10}Zr_7$ and Ni_3Zr. Such a mixture is consistent with the Ni-Zr phase diagram, given that the 2Ni|Zr composition lies in a dual-phase stability zone. The time-temperature profile during self-propagating reaction in the 2Ni|Zr foils was consistent with the occurrence of a rapid transformation which released up to 80% of the stored heat, and which was then followed by slow transformation into the final phases. It was assumed that the reaction pathways were similar in self-propagating reactions and in slow heating reactions of 2Ni|Zr, and that the rapid initial reaction was the formation of the metastable intermediate phase-field of two intermetallics: one being NiZr and the other an unidentified phase. Subsequent transformation into the above $Ni_{10}Zr_7$ and Ni_3Zr occurred during 40ms following the passage of the maximum-temperature isotherm.

Table 19. Maximum reaction temperature of Ni|Zr foils

Foil	Final Phase	$H_{reaction}$ (J/g)	Extent of Reaction (%)	T_{max} (K)
Ni|Zr	NiZr	477	71-75	1206
2Ni|Zr	$Ni_{10}Zr_2$, Ni_3Zr	518	70-82	1228
7Ni|2Zr	Ni_7Zr_2	518	71-80	1208

Materials Research Forum LLC
doi: http://dx.doi.org/10.21741/9781644900093

Miscellaneous

Metal\RE

Thin multilayers which comprised alternating scandium or yttrium, and silver, copper or gold layers, were deposited using direct-current magnetron sputtering[236]. The Sc\Au, Sc\Cu, Y\Au and Y\Cu multilayers reacted under vacuum to form single-phase cubic B2 structures (tables 20 and 21). Those multilayers which contained silver and a rare-earth formed mainly cubic B2 (RE)Ag plus some (RE)Ag$_2$. The propagation speeds ranged from 0.1 to 40.0m/s (table 22) and both steady-state and spin-like reaction modes occurred. Scandium- and yttrium-containing multilayers require reaction in vacuum in order to prevent oxidation.

Table 20 Enthalpies of rare-earth multilayer systems

System	Enthalpy (kJ/mol)
Sc\Ag	-26.2
Sc\Au	-76.1
Sc\Cu	-20.9
Y\Ag	-26.8
Y\Au	-78.7
Y\Cu	-19.3

Ti\B

Multilayers having periodicities of 50 to 3000nm and a Ti:B stoichiometry of 1:2 were prepared by using magnetron sputtering[237]. The as-deposited multilayers consisted of well-ordered layers of crystalline titanium and amorphous boron. The amount of intermixed material hardly affected the exothermicity. The propagation rates for foils with bilayer thicknesses of less than 666nm did not exhibit a pressure-dependence, while foils having bilayer thicknesses above 857nm did do so (figures 67 and 68).

Materials Research Forum LLC
doi: http://dx.doi.org/10.21741/9781644900093

Table 21. Estimated reaction temperatures in rare-earth multilayer systems

System	Compound Melting Point (K)	Reaction Temperature (K)
Sc|Ag	1503	1122
Sc|Au	1973	>1973
Sc|Cu	1399	994
Y|Ag	1433	1133
Y|Au	1873	>1873
Y|Cu	1220	935

These thick bilayer foils did not sustain propagation below a characteristic pressure. Post-reaction Auger electron spectroscopy revealed oxygen penetration into thicker bilayers, and incomplete mixing of the titanium and boron. The thinner bilayer foils were converted into single-phase hexagonal TiB_2. Increased air pressure promoted reaction propagation in thicker bilayers, and this was attributed to an increased heat release, due to oxidation, which promoted intermixing and reaction of titanium and boron. The reaction propagation rates ranged from 0.05 to 15m/s. The energies evolved by this system are discussed below in some detail.

Table 22. Propagation rates in rare-earth multilayer systems

System	Propagation Rate (m/s)
Sc|Ag	0.2 - 0.45
Sc|Au	8 - 40
Sc|Cu	0.4 - 0.9
Y|Ag	0.4 - 0.75
Y|Au	9 - 13
Y|Cu	0.18 - 0.4

The effect of the ambient gaseous environment upon the reaction behavior and product formation in sputter-deposited Ti|2B reactive multilayers was investigated by using air pressures ranging from atmospheric to 10^{-4}Torr. The reaction propagation rates for

3.0μm-thick multilayers ranged from 10.89 to 0.05m/s, depending upon the bilayer thickness and ambient pressure[238]. Single-phase TiB_2 formed in multilayers with a small bilayer thickness, while multilayers with a large bilayer thickness comprised a mixture of TiB_2, TiB and TiO_2. Numerical investigations have been made of reaction ignition and self-propagation in nanoscale multilayer structures made of alternating layers of boron and titanium[239]. Thanks to thinness of the metallic multilayers, the reaction propagation through the bimetallic multilayers could be approximated by a one-dimensional transient model for thermal and atomic diffusion. The reaction was assumed to obey an Arrhenius temperature-dependence. The results elucidated the roles played by the multilayer thickness, by an excess of one component and by the presence of pre-mixing at the interface between boron and titanium layers, in determining the ignition delay and the reaction propagation velocity.

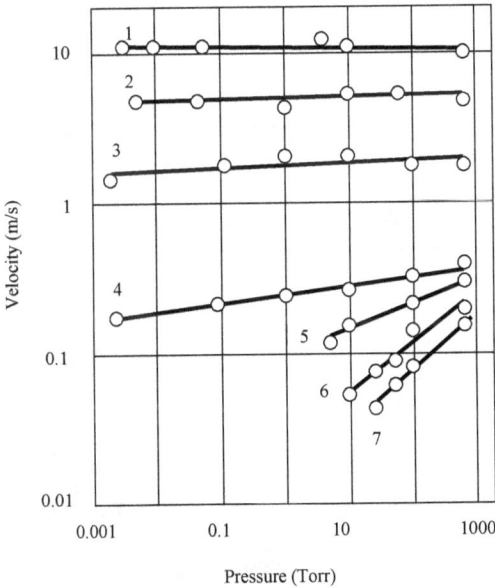

Figure 67. Propagation rate of reaction in Ti\2B foil as a function of air pressure and bilayer thickness: 1: 50nm, 2: 100nm, 3: 300nm, 4: 666nm, 5: 857nm, 6: 1225nm, 7: 3000nm

An integrated film initiator has been prepared by combining Ti|B nano-multilayers with a copper film bridge. The latter was first wet-etched and the Ti|B multilayers were then magnetron-sputtered onto the copper[240]. Self-propagating exothermic reaction of the 2μm-thick Ti|B multilayers could be initiated by 60V capacitor discharge. The reaction temperature could attain 2600K. Various reaction temperature could be obtained by varying the thickness of the Ti|B multilayers. As compared with a copper-film bridge, the present integrated film bridge offered a higher explosion temperature, a longer explosion duration, more violent behaviour and larger quantities of ejected product.

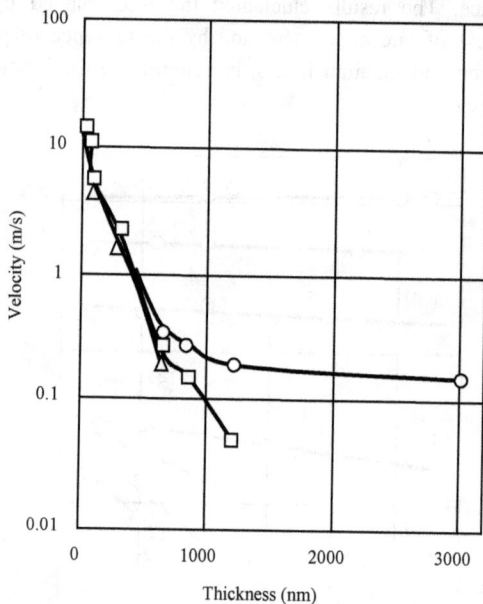

*Figure 68. Propagation rate of reaction in Ti|2B foil
as a function of bilayer thickness
Circles: 1torr, squares: 10torr, triangles: 0.1torr*

The overall reaction of titanium and boron to produce titanium diboride was analyzed under near-isobaric conditions[241]. The multi-component diffusion model used a generalized Fick's law formulation in which coefficients that were linked to the binary diffusivities were defined by Maxwell-Stefan relationships. An elementary depletion form, with Arrhenius temperature-dependent coefficients, was used to describe the reaction rate. An asymptotic analysis was applied in the limit of low strain-rates, where

Materials Research Forum LLC

doi: http://dx.doi.org/10.21741/9781644900093

there was a long residence-time in the reaction zone. A constant mixture density assumption simplified the flow description. The diffusion models assumed equal and unequal molecular weights for the various species. A full numerical study involved finite-rate chemistry, a composition-dependent density and strain-rates which ranged from low to moderate values. All of the approaches agreed very well in describing the flame structure, flame temperature and degree of incomplete combustion. There was found to exist a critical strain-rate, beyond which steady burning could no longer be observed.

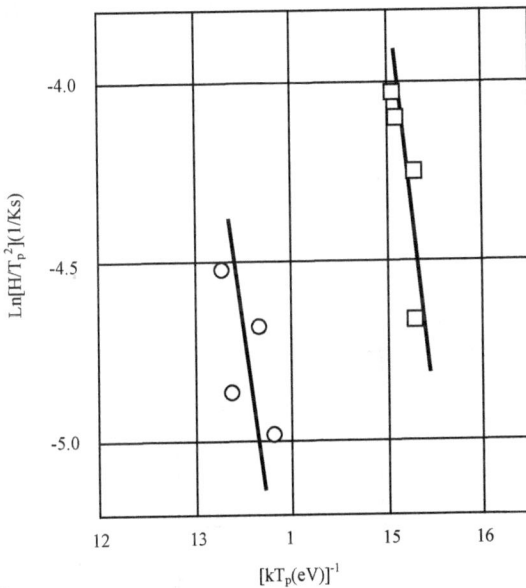

Figure 69. Kissinger plots for crystallization peaks
for 4.1nm (□) and 8.3nm (o) period Zr|B samples

Zr|B

Ultra-high temperature ZrB_2 coatings were prepared below 1300K by using Zr|B reactive multilayers. Highly-textured crystalline ZrB_2 could be formed at moderate temperatures because of an absence of oxide at the Zr|B interface and because of the very short diffusion distance. The Zr|B reaction occurred in two stages. In the first stage there was interdiffusion between the nanocrystalline zirconium and amorphous boron layers;

Materials Research Forum LLC
doi: http://dx.doi.org/10.21741/9781644900093

creating an amorphous Zr|B alloy[242]. In the second stage the amorphous alloy crystallized to form ZrB_2. Scanning nanocalorimetry, using heating rates ranging from 3100 to 10000K/s, revealed activation energies of 0.47 and 2.4eV for Zr|B interdiffusion and ZrB_2 crystallization, respectively. The activation energy for the crystallization process could be determined from a Kissinger plot (figure 69). The results for multilayers having an 8.3nm bilayer period exhibited significant scatter because the overlap of two peaks in the calorimetry traces made it difficult to determine the peak temperature accurately. The results for multilayers having a 4.1nm period were better and provided an estimate of the activation energy (2.4eV). Although it was not possible to deduce an accurate value for the activation energy from the data points for 8.3nm samples, the results were consistent with an activation energy of 2.4eV. A previously determined for the activation energy (1.69 to 1.99eV) of boron diffusion in zirconium was much higher than the diffusion activation energy found here.

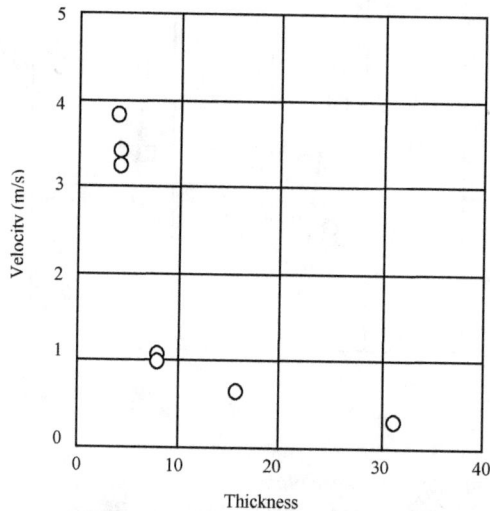

*Figure 70. Propagation velocity in Nb|Si multilayer
foils as a function of the bilayer thickness*

Nb|Si

Self-propagating reaction velocities in magnetron sputter-deposited 25μm Nb|Si nanoscale multilayer foils, having an average composition of 26at%Si, decreased as the individual niobium and silicon layers were thickened, and ranged overall from 0.3 to 3.8m/s[243]. The velocities nevertheless decreased smoothly from 4 to 0.5m/s (figure 70) as

Materials Research Forum LLC

doi: http://dx.doi.org/10.21741/9781644900093

the period of the niobium and silicon layers was increased from 39 to 312nm. Sputter-deposited 3Nb|Si reacted to yield a mixture of Nb_5Si_3, Nb and Nb_3Si; but predominantly niobium and Nb_5Si_3. The foils could be shaped during reaction. For example, when a circular foil was constrained on a glass substrate and then pressurized using the air enclosed by the foil and the glass, this forced the foil to deform plastically by as much as 15% at rates of up to 5/s.

Ni|Si

Explosive silicidation in nickel|amorphous-silicon multilayer thin films was investigated using a ratio of two nickel atoms to one silicon atom; yielding Ni_2Si. The temperature at the reaction front was estimated to be 1565K, while its velocity ranged from 22 to 27m/s. The velocity was very weakly dependent upon the modulation period and the total film thickness[244]. The resultant Ni_2Si grain structure was less defected than when produced by thermal annealing. This difference was attributed to the higher temperatures and shorter times which were involved in the former case. Differential scanning calorimetry was used to study the temperatures and kinetics involved in nickel silicide formation via the reaction of nickel|amorphous-silicon multilayer films[245]. When the layer thickness ratio of the film was 1:1, Ni_2Si was the only phase which was formed; a process which was described by:

$$D(cm^2/s) = 6 \times 10^0 exp[-1.5(eV)/kT]$$

The temperature at which Ni_2Si formation occurred was a function of layer thickness, with thinner layers reacting at lower temperatures. This thickness-dependence was attributed to a lower reaction time for thinner layers. Under mechanical impact, the films which comprised layers that were thinner than 125Å reacted explosively at room temperature to form Ni_2Si. Such rapid silicidation was suggested to occur when the rate of heat generation at the many reacting interfaces exceeded the rate of heat dissipation. Changing the individual layer thickness had a marked effect upon the temperature for Ni_2Si formation. As the layer thickness was reduced, Ni_2Si formed at lower temperatures. The peak maximum for 500Å nickel against 500Å amorphous silicon multilayers was 545K, while that for 70Å|70Å films was at 430K. A Kissinger analysis (figure 71) of 250Å|250Å film indicated that the activation energy for Ni_2Si formation was the same as for 500Å|500Å films. Kissinger[246] had shown that:

$$\ln(H/T_p^2) = C - E/kT_p$$

where H is the heating rate, T_p is the peak temperature, C is a constant and E is the activation energy of the reaction. The E-values for peaks 3 and 6 were 1.5 and 2.5eV,

Materials Research Forum LLC
doi: http://dx.doi.org/10.21741/9781644900093

respectively, and the activation energy for Ni_2Si formation agreed well with the previously reported value (1.4eV).

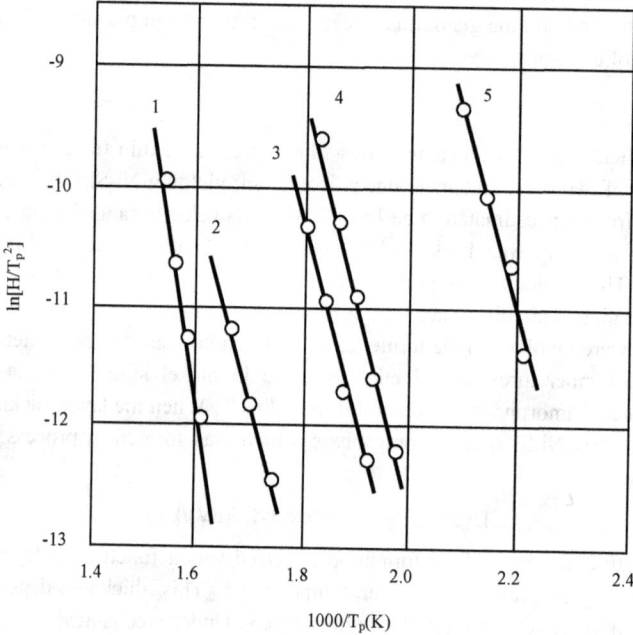

Figure 71. Kissinger plot of various differential scanning calorimetry peaks
(1: 2.5eV, 2: 1.4eV, 3: 1.5eV, 4: 1.5eV, 5: 1.4eV)
for 500Å nickel\500Å silicon multilayer films

Ti|Si

Exothermic reaction of these multilayers was used as an internal heat-source for bonding. Interaction of the very thin films generated sufficient heat to self-propagate the reaction and melt tin films[247]. The adhesion after the reaction front had passed depended strongly upon the substrate material. Self-propagation of the reaction in magnetron-sputtered nanoscale Ti|Si|Ti|nAl (n = 1, 2, 3) reactive multilayers was investigated as a function of the aluminium molar ratio in Ti|Al bilayers while keeping constant the Ti|Si bilayer and

Materials Research Forum LLC
doi: http://dx.doi.org/10.21741/9781644900093

total film thickness[248]. The reacted films of Ti|Si and Ti|Al bilayers contained Si_2Ti, $SiTi_3$ and Al_2Ti. The maximum reaction temperatures ranged from 1462 to 1812C during self-propagating reaction. The Ti|Si|Ti|Al and Ti|Si|Ti|2Al foils had reaction-front velocities of 9.1 and 7.7m/s, respectively. On the other hand, Ti|Si|Ti|3Al multilayer foils had the relatively low reaction temperature of 1462C and a low propagation velocity of 6.2m/s. A large amount of reaction heat was also generated by Ti|Si|Ti|Al multilayer foils. The three multilayer foils had the identical Ti|Si bilayer thickness of about 29nm and total thickness of 5 to 6µm, and so only the variation in Ti|Al bilayer thickness, from 29 to 47nm, caused differing local intermixing and reaction propagation rates. These results suggested that the reaction velocity, temperature and evolved heat of reactive multilayer foils could be optimized by adjusting the aluminium molar ratio. In a previous study, magnetron-sputtered reactive films having two different multilayer arrangements, Ti|Si|Ti|Al and Si|Ti|Al|Si, but with an identical bilayer thickness of 29nm and total thickness of 6µm, had been investigated[249]. The steady-state reaction front propagated through Ti|Si|Ti|Al films with a reaction velocity of 9.1m/s and a maximum reaction temperature of 1768C. Both steady-state and unsteady propagation was observed in Si|Ti|Al|Si foils, and this film had a maximum reaction velocity of 2.7m/s. The Ti|Si|Ti|Al film exhibited a higher reaction temperature and propagation velocity. Although both types of film involved similar reactants and morphologies, the differing reactant interfaces led to different interfacial reactions and local atomic mixing rates; thus affecting the overall reaction characteristics.

Rh|Si

Alternating layers of 400 to 500Å-thick silicon and 200Å-thick rhodium, adding up to 21 layers altogether, were electron-beam evaporated under high vacuum onto oxidized silicon substrates at room temperature[250]. Layer intermixing occurred, and the number of layers which was consumed during reaction was related to the degree of surface roughness. The presence of Rh_4Si_5 and Rh_3Si_4 was detected; both of which habitually formed via nucleation-controlled kinetics at above 800C. It was deduced that reaction had occurred via an explosive crystallization which had been initiated by hot silicon microparticles. Propagation of the crystallization wave-front around secondary silicon microparticles resulted in the formation of various geometric surface morphologies of three types; each related to the energy balance controlling propagation of the explosive crystallization wave-front.

Materials Research Forum LLC
doi: http://dx.doi.org/10.21741/9781644900093

Si|NaClO₄

Some energetic bonding formations are much closer in nature to true deflagratory explosives. Electronic components have been joined using this combination of porous silicon with sodium perchlorate[251]. Silicon channels, some 97μm wide and 45μm deep, have been etched, which burned at 4.6m/s, and even followed 90° changes in direction. The minimum spacing between porous silicon channels was investigated and it was shown that porous silicon channels, less than 40μm apart, could burn independently. Based upon this spacing, it was plausible to have path lengths which were longer than 50cm on a chip surface area of 1.65cm². Smaller porous silicon channels, 28μm wide and 14μm deep, were also possible but reacted less steadily.

References

[1] Fisher, D.J., Materials Research Foundations, 43, 2019. http://dx.doi.org/10.21741/9781644900055

[2] Makowiecki, D.M., Bionta, R.M., US Patent 5,381,944, 17th January 1995.

[3] Goldschmidt, H., DE96317, 13th March 1895.

[4] Merzhanov, A.G., Borovinskaya, I.P., Doklady Akademii Nauk SSSR, 204, 1972. 366.

[5] Gore, G., Philosophical Magazine, 9, 1855, 73. https://doi.org/10.1080/14786445508641829

[6] Coffin, C.C., Proceedings of the Royal Society A, 152, 1935, 47-63. https://doi.org/10.1098/rspa.1935.0178

[7] Myagkov, V.G., Bykova, L.E., Zhigalov, V.S., Matsynin, A.A., Volochaev, M.N., Tambasov, I.A., Mikhlin, Y.L., Bondarenko, G.N., Journal of Alloys and Compounds, 724, 2017, 820-826. https://doi.org/10.1016/j.jallcom.2017.07.081

[8] Rossi, C., Propellants, Explosives, Pyrotechnics, 2018, Article in Press.

[9] Blobaum, K.J., Reiss, M.E., Plitzko, M., Weihs, T.P., Journal of Applied Physics, 94, 2003, 2915-2922. https://doi.org/10.1063/1.1598296

[10] Manesh, N.A., Basu, S., Kumar, R., Combustion and Flame, 157[3] 2010, 476-480. https://doi.org/10.1016/j.combustflame.2009.07.011

[11] Petrantoni, M., Rossi, C., Salvagnac, L., Conédéra, V., Estèvel, A., Tenailleau, C., Alphonse, P., Chaba, Y.J., Journal of Applied Physics, 108[8] 2010, 084323. https://doi.org/10.1063/1.3498821

[12] Lanthony, C., Ducéré, J.M., Estève, A., Rossi, C., Djafari-Rouhani, M., Thin Solid Films, 520, 2012, 4768. https://doi.org/10.1016/j.tsf.2011.10.184

[13] Hu, B., Zhu, P., Shen, R., Ye, Y., Wu, L., Hu, Y., High Power Laser and Particle Beams, 27[2] 2015, 024108.

[14] Zhu, P., Shen, R., Ye, Y., Zhou, X., Hu, Y., Wu, L., Science and Technology of Energetic Materials, 73[5-6] 2012, 127-131.

[15] Zhu, P., Shen, R., Ye, Y., Zhou, X., Hu, Y., Journal of Applied Physics, 110[7] 2011, 074513. https://doi.org/10.1063/1.3646489

[16] Zhou, X., Shen, R., Ye, Y., Zhu, P., Hu, Y., Wu, L., Journal of Applied Physics, 110[9] 2011, 094505. https://doi.org/10.1063/1.3658617

[17] Lahiner, G., Nicollet, A., Zapata, J., Marín, L., Richard, N., Rouhani, M.D., Rossi, C., Estève, A., Journal of Applied Physics, 122[15] 2017, 155105. https://doi.org/10.1063/1.5000312

[18] Marín, L., Gao, Y., Vallet, M., Abdallah, I., Warot-Fonrose, B., Tenailleau, C., Lucero, A.T., Kim, J., Esteve, A., Chabal, Y.J., Rossi, C., Langmuir, 33[41] 2017, 11086-11093. https://doi.org/10.1021/acs.langmuir.7b02964

[19] Nicollet, A., Lahiner, G., Belisario, A., Souleille, S., Djafari-Rouhani, M., Estève, A., Rossi, C., Journal of Applied Physics, 121[3] 2017, 034503. https://doi.org/10.1063/1.4974288

[20] Zhu, P., Shen, R., Ye, Y., Fu, S., Li, D., Journal of Applied Physics, 113[18] 2013, 184505. https://doi.org/10.1063/1.4804315

[21] Blobaum, K.J., Wagner, A.J., Plitzko, J.M., Van Heerden, D., Fairbrother, D.H., Weihs, T.P., Journal of Applied Physics, 94, 2003, 2923. https://doi.org/10.1063/1.1598297

[22] Apperson, S., Shende, R.V., Subramanian, S., Tappmeyer, D., Gangopadhyay, S., Chen, K., Gangopadhyay, K., Redner, P., Nicholich, S., Kapoor, D., Applied Physics Letters, 91[24] 2007, 243109. https://doi.org/10.1063/1.2787972

[23] Zhanga, K., Rossia, C., Rodriguez, G.A.A., Tenailleau, C., Alphonse, P., Applied Physics Letters, 91[11] 2007, 113117. https://doi.org/10.1063/1.2785132

[24] Matsuda, T., Takahashi, M., Sano, T., Hirose, A., Materials and Design, 121, 2017, 136-142. https://doi.org/10.1016/j.matdes.2017.02.045

[25] Kinsey, A.H., Slusarski, K., Krumheuer, E., Weihs, T.P., Journal of Materials Science, 52[18] 2017, 11077-11090. https://doi.org/10.1007/s10853-017-1262-8

Materials Research Forum LLC
doi: http://dx.doi.org/10.21741/9781644900093

[26] Kinsey, A.H., Slusarski, K., Sosa, S., Weihs, T.P., ACS Applied Materials and Interfaces, 9[26] 2017, 22026-22036. https://doi.org/10.1021/acsami.7b03071

[27] Murray, A.K., Isik, T., Ortalan, V., Gunduz, I.E., Son, S.F., Chiu, G.Y.C., Rhoads, J.F., Journal of Applied Physics, 122[18] 2017, 184901. https://doi.org/10.1063/1.4999800

[28] Sullivan, K.T., Kuntz, J.D., Gash, A.E., Journal of Applied Physics, 112[2] 2012, 024316. https://doi.org/10.1063/1.4737464

[29] Xiong, G., Yang, C., Zhu, W., Applied Surface Science, 459, 2018, 835–844. https://doi.org/10.1016/j.apsusc.2018.08.069

[30] Kim, K.J., Cho, M.H., Kim, S.H., Combustion and Flame, 197, 2018, 319-327. https://doi.org/10.1016/j.combustflame.2018.08.016

[31] Tai, Y., Xu, J., Wang, F., Dai, J., Zhang, W., Ye, Y., Shen, R., Journal of Applied Physics, 123[23] 2018, 235302. https://doi.org/10.1063/1.5031068

[32] Xu, J., Tai, Y., Ru, C., Dai, J., Shen, Y., Ye, Y., Shen, R., Fu, S., Journal of Applied Physics, 121[11] 2017, 113301. https://doi.org/10.1063/1.4978371

[33] Xu, J., Tai, Y., Ru, C., Dai, J., Ye, Y., Shen, R., Zhu, P., ACS Applied Materials and Interfaces, 9[6] 2017, 5580-5589. https://doi.org/10.1021/acsami.6b14662

[34] Bockmon, B.S., Pantoya, M.L., Son, S.F., Asay, B.W., Mang, J.T., Journal of Applied Physics, 98[6] 2005, 064903. https://doi.org/10.1063/1.2058175

[35] Aureli, M., Doumanidis, C.C., Gunduz, I.E., Hussien, A.G.S., Liao, Y., Jaffar, S.M., Rebholz, C., Doumanidis, C.C., Journal of Applied Physics, 122[2] 2017, 025118. https://doi.org/10.1063/1.4993174

[36] Schreiber, S., Theodossiadis, G.D., Zaeh, M.F., IOP Conference Series - Materials Science and Engineering, 181[1] 2017, 012008. https://doi.org/10.1088/1757-899X/181/1/012008

[37] Shuck, C.E., Manukyan, K.V., Rouvimov, S., Rogachev, A.S., Mukasyan, A.S., Combustion and Flame, 163, 2016, 487-493. https://doi.org/10.1016/j.combustflame.2015.10.025

[38] Mukasyan, A.S., Shuck, C.E., Pauls, J.M., Manukyan, K.V., Moskovskikh, D.O., Rogachev, A.S., Advanced Engineering Materials, 20, 2018, 1701065. https://doi.org/10.1002/adem.201701065

[39] Mukasyan, A.S., Vadchenko, S.G., Khomenko, I.O., Combustion and Flame, 111[1-2] 1997, 65-72. https://doi.org/10.1016/S0010-2180(97)00095-3

Materials Research Forum LLC
doi: http://dx.doi.org/10.21741/9781644900093

[40] Strunina, A.G., Dvoryankin, A.V., Merzhanov, A.G., Fizika Goreniya i Vzryva, 19[2] 1983, 30-36.

[41] Jayaraman, S., Knio, O.M., Mann, A.B., Weihs, T.P., Journal of Applied Physics, 86[2] 1999, 800-809. https://doi.org/10.1063/1.370807

[42] Wickersham, C.E., Poole, J.E., Journal of Vacuum Science & Technology, 6, 1988, 1699. https://doi.org/10.1116/1.575315

[43] Wei, C.T., Nesterenko, V.F., Weihs, T.P., Remington, B.A., Park, H.S., Meyers, M.A., Acta Materialia, 60, 2012, 3929-3942. https://doi.org/10.1016/j.actamat.2012.03.028

[44] Peruško, D., Petrović, S., Kovač, J., Stojanović, Z., Panjan, M., Obradović, M., Milosavljević, M., Journal of Materials Science, 47, 2012, 4488. https://doi.org/10.1007/s10853-012-6311-8

[45] Peruško, D., Čizmović, M., Petrović, S., Siketić, Z., Mitrić, M., Pelicon, P., Dražić, G., Kovač, J., Milinović, V., Milosavljević, M., Laser Physics, 23[3] 2013, 036005. https://doi.org/10.1088/1054-660X/23/3/036005

[46] Picard, Y.N., McDonald, J.P., Friedmann, T.A., Yalisove, S.M., Adams, D.P., Applied Physics Letters, 93[10] 2008, 104104. https://doi.org/10.1063/1.2981570

[47] Armstrong, R., Combustion Science and Technology, 71, 1990, 155. https://doi.org/10.1080/00102209008951630

[48] Zaporozhets, T.V., Gusak, A.M., Ustinov, A.I., Advances in Electrometallurgy, 1, 2010, 40.

[49] Zaporozhets, T.V., Gusak, A.M., Ustinov, A.I., Advances in Electrometallurgy, 3, 2012, 38.

[50] Zaporozhets, T.V., Gusak, A.M., Korol, Y.D., Ustinov, A.I., International Journal of Self-Propagating High-Temperature Synthesis, 22, 2013, 222. https://doi.org/10.3103/S1061386213040092

[51] Amini-Manesh, N., Basu, S., Kumar, R., Energy, 36, 2011, 1688. https://doi.org/10.1016/j.energy.2010.12.061

[52] Weingarten, N.S., Rice, B.M., Journal of Physics - Condensed Matter, 27, 2011, 275701. https://doi.org/10.1088/0953-8984/23/27/275701

[53] Weingarten, N.S., Mattson, W.D., Yau, A.D., Weihs, T.P, Rice, B.M., Journal of Applied Physics, 107[9] 2010, 093517. https://doi.org/10.1063/1.3340965

Materials Research Forum LLC
doi: http://dx.doi.org/10.21741/9781644900093

[54] Gunduz, I.E., Fadenberger, K., Kokonou, M., Rebholz, C., Doumanidis, C.C., Ando, T., Journal of Applied Physics, 105[7] 2009, 074903. https://doi.org/10.1063/1.3091284

[55] Salloum, M., Knio, O.M., Combustion and Flame, 157, 2010, 436. https://doi.org/10.1016/j.combustflame.2009.08.010

[56] Salloum, M., Knio, O.M., Combustion and Flame, 157, 2010, 1154. https://doi.org/10.1016/j.combustflame.2009.10.005

[57] Salloum, M., Knio, O.M., Combustion and Flame, 157[2] 2010, 288-295. https://doi.org/10.1016/j.combustflame.2009.06.019

[58] Jayaraman, S., Knio, O.M., Mann, A.B., Weihs, T.P., Journal of Applied Physics, 86, 1999, 800. https://doi.org/10.1063/1.370807

[59] Khina, B.B., Journal of Applied Physics, 101[6] 2007, 063510. https://doi.org/10.1063/1.2710443

[60] Trenkle, J.C., Wang, J., Weihs, T.P., Hufnagel, T.C., Applied Physics Letters, 87[15] 2005, 153108. https://doi.org/10.1063/1.2099544

[61] Trenkle, J.C., Koerner, L.J., Tate, M.W., Gruner, S.M., Weihs, T.P., Hufnagel, T.C., Applied Physics Letters, 93[8] 2008, 081903. https://doi.org/10.1063/1.2975830

[62] Rogachev, A.S., Vadchenko, S.G., Mukasyan, A.S., Applied Physics Letters, 101[6] 2012, 063119. https://doi.org/10.1063/1.4745201

[63] Joress, H., Barron, S.C., Livi, K.J.T., Aronhime, N., Weihs, T.P., Applied Physics Letters, 101[11] 2012, 111908. https://doi.org/10.1063/1.4752133

[64] Raić, K.T., Rudolf, R., Kosec, B., Anžel, I., Lazić, V., Todorović, A., Materiali in Tehnologije, 43[1] 2010, 3-9.

[65] Reeves, R.V., Adams, D.P., Journal of Applied Physics, 115[4] 2014, 044911. https://doi.org/10.1063/1.4863339

[66] Adams, D.P., Hodges, V.C., Bai, M.M., Jones, E., Rodriguez, M.A., Buchheit, T., Moore, J.J., Journal of Applied Physics, 104[4] 2008, 043502. https://doi.org/10.1063/1.2968444

[67] Dreizin, E.L., Schoenitz, M., Progress in Energy and Combustion Science, 50, 2015, 81-105. https://doi.org/10.1016/j.pecs.2015.06.001

[68] Adams, D.P., Hodges, V.C., Bai, M.M., Jones, E., Rodriguez, M.A., Buchheit, T., Moore, J.J., Journal of Applied Physics, 104[4] 2008, 043502. https://doi.org/10.1063/1.2968444

[69] McDonald, J.P., Hodges, V.C., Jones, R.D., Adams, D.P., Applied Physics Letters, 94[3] 2009, 034102. https://doi.org/10.1063/1.3070119

[70] Kim J.S., LaGrange, T., Reed, B.W., R.Knepper, R., Weihs, T.P., Browning, N.D., Campbell, G.H., Acta Materialia, 59[9] 2011, 3571-3580. https://doi.org/10.1016/j.actamat.2011.02.030

[71] Dyer, T.S., Munir, Z.A., Metallurgical and Materials Transactions B, 26[3] 1995, 603-610. https://doi.org/10.1007/BF02653881

[72] Ma, E., Thompson, C.V., Clevenger, A., Tu, K.N., Applied Physics Letters, 57[12] 1990, 1262-1264. https://doi.org/10.1063/1.103504

[73] Trenkle, J.C., Koerner, L.J., Tate, M.W., Walker, N., Gruner, S.M., Weihs, T.P., Hufnagel, T.C., Journal of Applied Physics 107[11] 2010, 113511. https://doi.org/10.1063/1.3428471

[74] Knepper, R., Snyder, M.R., Fritz, G., Fisher, K., Knio, O.M., Weihs, T.P., Journal of Applied Physics, 105[8] 2009, 083504. https://doi.org/10.1063/1.3087490

[75] Gavens, A.J., Van Heerden, D., Mann, A.B., Reiss, M.E., Weihs, T.P., Journal of Applied Physics, 87[3] 2000, 1255-1263. https://doi.org/10.1063/1.372005

[76] McDonald, J.P., Reeves, R.V., Jones, E.D., Chinn, K.A., Adams, D.P., Journal of Applied Physics, 113[10] 2013, 103505. https://doi.org/10.1063/1.4794183

[77] Zhao, S., Germann, T.C., Strachan, A., The Journal of Chemical Physics, 125[16] 2006, 164707. https://doi.org/10.1063/1.2359438

[78] Xu, R.G., Falk, M.L., Weihs, T.P., Journal of Applied Physics, 114[16] 2013, 163511. https://doi.org/10.1063/1.4826527

[79] Grapes, M.D., Woll, K., Barron, D.C., LaVan, D.A., Weihs, T.P., Journal of Applied Physics, 113[14] 2013, 143509. https://doi.org/10.1063/1.4799628

[80] Sun, Y., Lu, Q., Song, G., Yong, Z., He, X., Journal of Functional Materials, 47[11] 2016, 11027-11033.

[81] Wang, Y., Wang, M., Guo, F., Fu, Q.B., Journal of Beijing Institute of Technology, 25, 2016, 231-235.

[82] Grapes, M.D., LaGrange, T., Woll, K., Reed, B.W., Campbell, G.H., LaVan, D.A., Weihs, T.P., APL Materials, 2[11] 2014, 116102. https://doi.org/10.1063/1.4900818

[83] Baras, F., Turlo, V., Politano, O., Journal of Materials Engineering and Performance, 25[8] 2016, 3270-3274. https://doi.org/10.1007/s11665-016-1989-4

[84] Alawieh, L., Knio, O.M., Weihs, T.P., Journal of Applied Physics, 110[1] 2011, 013509. https://doi.org/10.1063/1.3599847

[85] Schumacher, A., Gais, U., Knappmann, S., Dietrich, G., Braun, S., Pflug, E., Roscher, F., Vogel, K., Hertel, S., Kahler, D., Reinert, W., 20th European Microelectronics and Packaging Conference and Exhibition, 2016, 7390707.

[86] Qiu, X., Wang, J., Materials Research Society Symposium Proceedings, 968, 2007, 51-56.

[87] Qiu, X., Wang, J., Sensors and Actuators A, 141[2] 2008, 476-481. https://doi.org/10.1016/j.sna.2007.10.039

[88] Gibbins, J.D., Stover, A.K., Krywopusk, N.M., Woll, K., Weihs, T.P., Combustion and Flame, 162[12] 2015, 4408-4416. https://doi.org/10.1016/j.combustflame.2015.08.003

[89] Lu, H., Yuan, X., Cui, W., Ning, X., Chen, K., Key Engineering Materials, 519, 2012, 104-107. https://doi.org/10.4028/www.scientific.net/KEM.519.104

[90] Wang, L., He, B., Jiang, X.H., Fu, Q.B., Wang, L.L., Chinese Journal of Energetic Materials, 17[2] 2009, 233-235.

[91] Turlo, V., Politano, O., Baras, F., Journal of Applied Physics, 121[5] 2017, 055304. https://doi.org/10.1063/1.4975474

[92] Wang, T., Zeng, Q., Li, M., Central European Journal of Energetic Materials, 14[3] 2017, 547-558. https://doi.org/10.22211/cejem/75605

[93] Grapes, M.D., Weihs, T.P., Combustion and Flame, 172, 2016, 105-115. https://doi.org/10.1016/j.combustflame.2016.07.006

[94] Manukyan, K.V., Tan, W., Deboer, R.J., Stech, E.J., Aprahamian, A., Wiescher, M., Rouvimov, S., Overdeep, K.R., Shuck, C.E., Weihs, T.P., Mukasyan, A.S., ACS Applied Materials and Interfaces, 7[21] 2015, 11272-11279. https://doi.org/10.1021/acsami.5b01415

[95] Overdeep, K.R., Weihs, T.P., Journal of Thermal Analysis and Calorimetry, 122[2] 2015, 787-794. https://doi.org/10.1007/s10973-015-4805-8

[96] Spies, I., Schumacher, A., Knappmann, S., Rheingans, B., Janczak-Rusch, J., Jeurgens, L.P.H., 21st European Microelectronics and Packaging Conference and Exhibition, 2018, 1-6.

[97] Baras, F., Politano, O., Acta Materialia, 148, 2018, 133-146. https://doi.org/10.1016/j.actamat.2018.01.035

[98] Alawieh, L., Weihs, T.P., Knio, O.M., Combustion Theory and Modelling, 19[3] 2015, 329-346. https://doi.org/10.1080/13647830.2015.1026400

[99] Rogachev, A.S., Vadchenko, S.G., Baras, F., Politano, O., Rouvimov, S., Sachkova, N.V., Mukasyan, A.S., Acta Materialia, 66, 2014, 86-96. https://doi.org/10.1016/j.actamat.2013.11.045

[100] Braeuer, J., Besser, J., Wiemer, M., Gessner, T., Sensors and Actuators A, 188, 2012, 212-219. https://doi.org/10.1016/j.sna.2012.01.015

[101] Simões, S., Viana, F., Koak, M., Ramos, A.S., Vieira, M.T., Vieira, M.F., Journal of Materials Engineering and Performance, 21[5] 2012, 678-682. https://doi.org/10.1007/s11665-012-0144-0

[102] Ramos, A.S., Vieira, M.T., Simões, S., Viana, F., Vieira, M.F., Defect and Diffusion Forum, 297-301, 2010, 972-977.

[103] Simões, S., Viana, F., Ramos, A.S., Vieira, M.T., Vieira, M.F., Microscopy and Microanalysis, 16[6] 2010, 662-669. https://doi.org/10.1017/S143192761009392X

[104] Simões, S., Viana, F., Koak, M., Ramos, A.S., Vieira, M.T., Vieira, M.F., Materials Chemistry and Physics, 128[1-2] 2011, 202-207. https://doi.org/10.1016/j.matchemphys.2011.02.059

[105] Ramos, A.S., Vieira, M.T., Simões, S., Viana, F., Vieira, M.F., Advanced Materials Research, 59, 2009, 225-229.

[106] Crone, J.C., Knap, J., Chung, P.W., Rice, B.M., Proceedings of the 14th International Detonation Symposium, 2010, 13-22.

[107] Crone, J.C., Knap, J., Chung, P.W., Rice, B.M., Applied Physics Letters, 98[14] 2011, 141910. https://doi.org/10.1063/1.3575576

[108] Kecskes, L.J., Roberts, A.J., Mathaudhu, S.N., Klotz, B.R., Qiu, X., Graeter, J., Wang, J., Proceedings of the World Congress on Powder Metallurgy and Particulate Materials 2008, 990-1103.

[109] Qiu, X., Graeter, J., Kecskes, L., Wang, J., Journal of Materials Research, 23[2] 2008, 367-375. https://doi.org/10.1557/JMR.2008.0043

[110] Qiu, X., Graeter, J.H., Kecskes, L., Wang, J., Materials Research Society Symposium Proceedings, 977, 2006, 115-119.

[111] Beason, M.T., Gunduz, I.E., Son, S.F., Acta Materialia, 133, 2017, 247-257. https://doi.org/10.1016/j.actamat.2017.05.042

[112] Justice, A., Gunduz, I.E., Son, S.F., Thin Solid Films, 620, 2016, 48-53. https://doi.org/10.1016/j.tsf.2016.07.090

[113] Bridges, D., Rouleau, C., Gosser, Z., Smith, C., Zhang, Z., Hong, K., Cheng, J., Bar-Cohen, Y., Hu, A., Applied Sciences, 8[6] 2018, 985. Slightly modified and rearranged figures reproduced from doi:10.3390/app8060985 under Creative Commons license. https://doi.org/10.3390/app8060985

[114] Maj, Ł., Morgiel, J., Mars, K., Tarasek, A., Godlewska, E., Metallurgical and Materials Transactions A, 49[11] 2018, 5423-5427. https://doi.org/10.1007/s11661-018-4843-5

[115] Maj, Ł., Morgiel, J., Mars, K., Grzegorek, J., Faryna, M., Godlewska, E., Journal of Materials Processing Technology, 255, 2018, 689-695. https://doi.org/10.1016/j.jmatprotec.2018.01.023

[116] Feng, G., Li, Z., Jacob, R.J., Yang, Y., Wang, Y., Zhou, Z., Sekulic, D.P., Zachariah, M.R., Materials and Design, 126, 2017, 197-206. https://doi.org/10.1016/j.matdes.2017.04.044

[117] Rogachev, A.S., Vadchenko, S.G., Aronin, A.S., Shchukin, A.S., Kovalev, D.Y., Nepapushev, A.A., Rouvimov, S., Mukasyan, A.S., Journal of Alloys and Compounds, 749, 2018, 44-51. https://doi.org/10.1016/j.jallcom.2018.03.255

[118] Rogachev, A.S., Vadchenko, S.G., Shchukin, A.S., International Journal of Self-Propagating High-Temperature Synthesis, 26[1] 2017, 44-48. https://doi.org/10.3103/S1061386217010095

[119] Rogachev, A.S., Vadchenko, S.G., Baras, F., Politano, O., Rouvimov, S., Sachkova, N.V., Grapes, M.D., Weihs, T.P., Mukasyan, A.S., Combustion and Flame, 166, 2016, 158-169. https://doi.org/10.1016/j.combustflame.2016.01.014

[120] Ramos, A.S., Maj, L., Morgiel, J., Vieira, M.T., Metals, 7[12] 2017, 574. https://doi.org/10.3390/met7120574

[121] Oh, M., Oh, M.C., Han, D., Jung, S.H., Ahn, B., Metals, 8[2] 2018, 121. https://doi.org/10.3390/met8020121

[122] Schnabel, V., Sologubenko, A.S., Danzi, S., Kurtuldu, G., Spolenak, R., Applied Physics Letters, 111[17] 2017, 173902. https://doi.org/10.1063/1.5003219

[123] Zhao, H., Tan, C., Yu, X., Ning, X., Nie, Z., Cai, H., Wang, F., Cui, Y., Journal of Alloys and Compounds, 741, 2018, 883-894. https://doi.org/10.1016/j.jallcom.2018.01.170

Materials Research Forum LLC
doi: http://dx.doi.org/10.21741/9781644900093

[124] Fritz, G.M., Grzyb, J.A., Knio, O.M., Grapes, M.D., Weihs, T.P., Journal of Applied Physics, 118[13] 2015, 135101. https://doi.org/10.1063/1.4931666

[125] Lu, S., Mily, E.J., Irving, D.L., Maria, J.P., Brenner, D.W., Journal of Physical Chemistry C, 119[25] 2015, 14411-14418.

[126] Hooper, R.J., Adams, D.P., Hirschfeld, D., Manuel, M.V., Journal of Electronic Materials, 45[1] 2016, 1-11. https://doi.org/10.1007/s11664-015-3941-z

[127] Hooper, R.J., Davis, C.G., Johns, P.M., Adams, D.P., Hirschfeld, D., Nino, J.C., Manuel, M.V., Journal of Applied Physics, 117[24] 2015, 245104. https://doi.org/10.1063/1.4922981

[128] Grieseler, R., Welker, T., Müller, J., Schaaf, P., Physica Status Solidi A, 209[3] 2012, 512-518. https://doi.org/10.1002/pssa.201127470

[129] Gunduz, I.E., Onel, S., Doumanidis, C.C., Rebholz, C., Son, S.F., Journal of Applied Physics, 117[21] 2015, 214904. https://doi.org/10.1063/1.4921906

[130] Wołczyński, W., Archives of Metallurgy and Materials, 60[3] 2015, 2421-2429. https://doi.org/10.1515/amm-2015-0395

[131] Gunduz, I.E., Fadenberger, K., Kokonou, M., Rebholz, C., Doumanidis, C.C., Ando, T., Journal of Applied Physics, 105[7] 2009, 074903. https://doi.org/10.1063/1.3091284

[132] Gust, W., Hintz, M.B., Lodding, A., Odelius, H., Predel, B., Physica Status Solidi A, 64[1] 1981, 187-194. https://doi.org/10.1002/pssa.2210640120

[133] Yakoviev, S.G., Mozzhukin, E.I., Smirnov, P.B., Izvest. V.U.Z. Tsvetnaya Met., 4, 1974, 145-148.

[134] Sivakumar, R., Seigle, L.L., Metallurgical Transactions A, 7[8] 1976, 1073-1080. https://doi.org/10.1007/BF02656589

[135] Berndt, K., Karin, H., Crystal Research and Technology, 20[11] 1985, 1461-1465. https://doi.org/10.1002/crat.2170201107

[136] Tonejc, A., Philosophical Magazine, 27[3] 1973, 753-755. https://doi.org/10.1080/14786437308219245

[137] Grammatikakis, J., Eftaxias, K., Hadjicontis, V., Journal of the Physics and Chemistry of Solids, 49[10] 1988, 1275-1277. https://doi.org/10.1016/0022-3697(88)90186-2

[138] Barinov, G.I., Tekhnol. Mater. Elektron. Tekh., 1970, 82-86.

Materials Research Forum LLC
doi: http://dx.doi.org/10.21741/9781644900093

[139] Barinov, G.I., Tekhnol. Mater. Elektron. Tekh., 1970, 82-86.

[140] Wei, H., Hou, G.C., Zheng, Q., Sun, X.F., Yao, X.D., Guan, H.R., Hu, Z.Q., Journal of Physics - Conference Series, 96[1] 2008, 012183. https://doi.org/10.1088/1742-6596/96/1/012183

[141] Watanabe, M., Horita, Z., Sano, T., Nemoto, M., Acta Metallurgica et Materialia, 42[10] 1994, 3389-3396. https://doi.org/10.1016/0956-7151(94)90471-5

[142] Damoc, L., Fonda, E., Le Fevre, P., Traverse, A., Journal of Applied Physics, 92[4] 2002, 1862-1867. https://doi.org/10.1063/1.1493652

[143] Nakamura, R., Takasawa, K., Yamazaki, Y., Iijima, Y., Intermetallics, 10[2] 2002, 195-204. https://doi.org/10.1016/S0966-9795(01)00125-X

[144] Poletaev, G.M., Starostenkov, M.D., Technical Physics Letters, 29[6] 2003, 454-455. https://doi.org/10.1134/1.1589555

[145] Ren, X., Chen, G.Q., Zhou, W.L., Wu, C.W., Zhang, J.S., Journal of Alloys and Compounds, 472[1-2] 2009, 525-529. https://doi.org/10.1016/j.jallcom.2008.05.010

[146] Mohanty, R.R., Leon, A., Sohn, Y.H., Computational Materials Science, 43[2] 2008, 301-308. https://doi.org/10.1016/j.commatsci.2007.11.002

[147] Wang, J., Besnoin, E., Duckham, A., Spey, S.J., Reiss, M.E., Knio, O.M., Powers, M., Whitener, M., Weihs, T.P., Applied Physics Letters, 83[19] 2003, 3987-3989. https://doi.org/10.1063/1.1623943

[148] Wang, J., Besnoin, E., Knio, O.M., Weihs, T.P., Acta Materialia, 52[18] 2004, 5265-5274. https://doi.org/10.1016/j.actamat.2004.07.012

[149] Wang, J., Besnoin, E., Knio, O.M., Weihs, T.P., Journal of Applied Physics, 97[11] 2005, 114307. https://doi.org/10.1063/1.1915540

[150] Swiston, A.J., Hufnagel, T.C., Weihs, T.P., Scripta Materialia, 48[12] 2003, 1575-1580. https://doi.org/10.1016/S1359-6462(03)00164-7

[151] Trenkle, J.C., Weihs, T.P., Hufnagel, T.C., Scripta Materialia, 58[4] 2008, 315-318. https://doi.org/10.1016/j.scriptamat.2007.09.060

[152] A.J.Swiston, Besnoin, E., Duckham, A., Knio, O.M., Weihs, T.P., Hufnagel, T.C., Acta Materialia, 53[13] 2005, 3713-3719. https://doi.org/10.1016/j.actamat.2005.04.030

[153] Swiston, A.J., Weihs, T.P., Knio, O.M., Hufnagel, T.C., Materials Research Society Symposium Proceedings, 806, 2003, 121-126.

Materials Research Forum LLC
doi: http://dx.doi.org/10.21741/9781644900093

[154] Fadenberger, K., Gunduz, I.E., Tsotsos, C., Kokonou, M., Gravani, S., Brandstetter, S., Bergamaschi, A., Schmitt, B., Mayrhofer, P.H., Doumanidis, C.C., Rebholz, C., Applied Physics Letters, 97[14] 2010, 144101. https://doi.org/10.1063/1.3485673

[155] Boettge, B., Braeuer, J., Wiemer, M., Petzold, M., Bagdahn, J., Gessner, T., Journal of Micromechanics and Microengineering, 20[6] 2010, 064018. https://doi.org/10.1088/0960-1317/20/6/064018

[156] Mann, A.B., Gavens, A.J., Reiss, M.E., Van Heerden, D., Bao, G., Weihs, T.P., Journal of Applied Physics, 82, 1997, 1178-1188. https://doi.org/10.1063/1.365886

[157] Jayaraman, S., Mann, A.B., Reiss, M., Weihs, T.P., Knio, O.M., Combustion and Flame, 124[1-2] 2001, 178-194. https://doi.org/10.1016/S0010-2180(00)00192-9

[158] Besnoin, E., Cerutti, S., Knio, O.M., Journal of Applied Physics, 92, 2002, 5474. https://doi.org/10.1063/1.1509840

[159] Simões, S., Viana, F., Vieira, M.F., Journal of Materials Engineering and Performance, 23[5] 2014, 1536-1543. https://doi.org/10.1007/s11665-014-0923-x

[160] Bezpalchuk, V.M., Marchenko, S.V., Rymar, O.M., Bogatyryov, O.O., Gusak, A.M., Metallofizika i Noveishie Tekhnologii, 37[1] 2015, 87-102. https://doi.org/10.15407/mfint.37.01.0087

[161] Maj, Ł., Morgiel, J., Thin Solid Films, 621, 2017, 165-170. https://doi.org/10.1016/j.tsf.2016.12.004

[162] Theodossiadis, G.D., Zaeh, M.F., Production Engineering, 11[3] 2017, 245-253. https://doi.org/10.1007/s11740-017-0733-8

[163] Theodossiadis, G.D., Zaeh, M.F., Production Engineering, 11[4-5] 2017, 373-381. https://doi.org/10.1007/s11740-017-0738-3

[164] Theodossiadis, G.D., Zaeh, M.F., Production Engineering, 11[4-5] 2017, 401-408. https://doi.org/10.1007/s11740-017-0753-4

[165] Simões, S., Ramos, A.S., Viana, F., Emadinia, O., Vieira, M.T., Vieira, M.F., Journal of Materials Engineering and Performance, 25[10] 2016, 4394-4401. https://doi.org/10.1007/s11665-016-2304-0

[166] Duckham, A., Brown, M., Besnoin, E., Van Heerden, D., Knio, O.M., Weihs, T.P., Ceramic Engineering and Science Proceedings, 25[3] 2004, 597-603. https://doi.org/10.1002/9780470291184.ch87

[167] Fritz, G.M., Spey, S.J., Grapes, M.D., Weihs, T.P., Journal of Applied Physics, 113[1] 2013, 014901. https://doi.org/10.1063/1.4770478

[168] Braeuer, J., Besser, J., Tomoscheit, E., Klimm, D., Anbumani, S., Wiemer, M., Gessner, T., ECS Transactions, 50, 2012, 241. https://doi.org/10.1149/05007.0241ecst

[169] Braeuer, J., Gessner, T., Journal of Micromechanics and Microengineering, 24[11] 2014, 115002. https://doi.org/10.1088/0960-1317/24/11/115002

[170] Braeuer, J., Besser, J., Wiemer, M., Gessner, T., 4th Electronic System-Integration Technology Conference, 2012, 6542144.

[171] Masser, R., Braeuer, J., Gessner, T., Journal of Applied Physics, 115[24] 2014, 244311. https://doi.org/10.1063/1.4885457

[172] Matkoviç, T., Schubert, K., Journal of Less Common Metals, 55, 1977, 45. https://doi.org/10.1016/0022-5088(77)90258-2

[173] Altunin, R.R., Moiseenko, E.T., Zharkov, S.M., Physics of the Solid State, 60[7] 2018, 1413-1418. https://doi.org/10.1134/S106378341807003X

[174] Abere, M.J., Yarrington, C.D., Adams, D.P., Journal of Applied Physics, 123[23] 2018, 235304. https://doi.org/10.1063/1.5026507

[175] Kittell, D.E., Yarrington, C.D., Hobbs, M.L., Abere, M.J., Adams, D.P., Journal of Applied Physics, 123[14] 2018, 145302. https://doi.org/10.1063/1.5025820

[176] Yarrington, C.D., Abere, M.J., Adams, D.P., Hobbs, M.L., Journal of Applied Physics, 121[13] 2017, 134301. https://doi.org/10.1063/1.4979578

[177] McDonald, J.P., Picard, Y.N., Yalisove, S.M., Adams, D.P., Optics InfoBase Conference Papers, 2009.

[178] McDonald, J.P., Picard, Y.N., Yalisove, S.M., Adams, D.P., Materials Research Society Symposium Proceedings, 1146, 2008, 135-139.

[179] Adams, D.P., Rodriguez, M.A., Tigges, C.P., Kotula, P.G., Journal of Materials Research, 21[12] 2006, 3168-3179. https://doi.org/10.1557/jmr.2006.0387

[180] Picard, Y.N., Adams, D.P., Palmer, J.A., Yalisove, S.M., Applied Physics Letters, 88[14] 2006, 144102. https://doi.org/10.1063/1.2191952

[181] Murphy, R.D., Reeves, R.V., Yarrington, C.D., Adams, D.P., Applied Physics Letters, 107[23] 2015, 234103. https://doi.org/10.1063/1.4937161

[182] Chang, C.A, Journal of Applied Physics, 52[7] 1981, 4620-4622. https://doi.org/10.1063/1.329341

[183] Bergner, D., Schwarz, K., Neue Hütte, 23[6] 1978, 210-212.

Materials Research Forum LLC

doi: http://dx.doi.org/10.21741/9781644900093

[184] Adams, D.P., Reeves, R.V., Abere, M.J., Sobczak, C., Yarrington, C.D., Rodriguez, M.A., Kotula, P.G., Journal of Applied Physics, 124[9] 2018, 095105. https://doi.org/10.1063/1.5026293

[185] Woll, K., Bergamaschi, A., Avchachov, K., Djurabekova, F., Gier, S., Pauly, C., Leibenguth, P., Wagner, C., Nordlund, K., Mücklich, F., Scientific Reports, 6, 2016, 19535. https://doi.org/10.1038/srep19535

[186] Woll, K., Gunduz, I.E., Pauly, C., Doumanidis, C.C., Son, S.F., Rebholz, C., Mücklich, F., Applied Physics Letters, 107[7] 2015, 073103. https://doi.org/10.1063/1.4928665

[187] Rogachev, A.S., Grigoryan, A.E., Illarionova, E.V., Kanel, I.G., Merzhanov, A.G., Nosyrev, A.N., Sachkova, N.V., Khvesyuk, V.I., Tsygankov, P.A., Combustion, Explosion and Shock Waves, 40[2] 2004, 166–171. https://doi.org/10.1023/B:CESW.0000020138.58228.65

[188] Gachon, J.C., Rogachev, A.S., Grigory, H.E., Illarionova, E.V., Kuntz, J.J., Kovalev, D.Y., Nosyrev, A.N., Sachkova, N.V., Tsygankov, P.A., Acta Materialia, 53[4] 2005, 1225-1231. https://doi.org/10.1016/j.actamat.2004.11.016

[189] Yi, J., Zhang, Y., Wang, X., Dong, C., Hu, H., Materials Transactions, 57[9] 2016, 1494-1497. https://doi.org/10.2320/matertrans.M2016126

[190] Sen, S., Lake, M., Wilden, J., Schaaf, P., Thin Solid Films, 631, 2017, 99-105. https://doi.org/10.1016/j.tsf.2017.04.012

[191] An, R., Tian, Y., Kong, L., Wang, C., Chang, S., Acta Metallurgica Sinica, 50[8] 2014, 937-943.

[192] Wang, L., He, B., Jiang, X.H., Combustion Science and Technology, 182[8] 2010, 1000-1008. https://doi.org/10.1080/00102200903489311

[193] Gaković, B., Tsibidis, G.D., Skoulas, E., Petrović, S.M., Vasić, B., Stratakis, E., Journal of Applied Physics, 122[22] 2017, 223106. https://doi.org/10.1063/1.5016548

[194] Sen, S., Lake, M., Kroppen, N., Farber, P., Wilden, J., Schaaf, P., Applied Surface Science, 396, 2017, 1490-1498. https://doi.org/10.1016/j.apsusc.2016.11.197

[195] Cao, J., Feng, J.C., Li, Z.R., Transactions of the China Welding Institution, 26[11] 2005, 5-7.

[196] Araki, H., Yamane, T., Minamino, Y., Saji, S., Hana, Y., Jung, S.B., Metallurgical Transactions A, 25[4] 1994, 874-876. https://doi.org/10.1007/BF02665465

[197] Tarento, R.J., Blaiser, G., Acta Metallurgica, 37[9] 1989, 2305-2312. https://doi.org/10.1016/0001-6160(89)90027-8

[198] Köppers, M., Herzig, C., Friesel, M., Mishin, Y., Acta Materialia, 45[10] 1997, 4181-4191. https://doi.org/10.1016/S1359-6454(97)00078-5

[199] Araujo, L.L., Behar, M., Applied Physics A, 71[2] 2000, 169-174.

[200] Nonaka, K., Fujii, H., Nakajima, H., Materials Transactions, 42[8] 2001, 1731-1740. https://doi.org/10.2320/matertrans.42.1731

[201] Räisänen, J., Anttila, A., Keinonen, J., Journal of Applied Physics, 57[2] 1985, 613-614. https://doi.org/10.1063/1.334747

[202] Xu, L., Cui, Y.Y., Hao, Y.L., Yang, R., Materials Science and Engineering A, 435-436, 2006, 638-647. https://doi.org/10.1016/j.msea.2006.07.077

[203] Romankov, S.E., Mamaeva, A., Vdovichenko, E., Ermakov, E., Nuclear Instruments and Methods in Physics Research B, 237[3-4] 2005, 575-584. https://doi.org/10.1016/j.nimb.2005.02.020

[204] Gershinskii, A.E., Fiz. Met. Metalloved, 32[5] 1971, 1104-1107.

[205] Gershinskii, A.E., Kostov, E.G., Fiz. Met. Metalloved, 30[6] 1970, 1315-1317.

[206] Nahar, R.K., Devashrayee, N.M., Khokle, W.S., Journal of Vacuum Science and Technology B, 6[3] 1988, 880-883. https://doi.org/10.1116/1.584315

[207] Pokoev, A.V., Mironov, V.M., Kudryavtseva, L.K., Izv. Vyssh. Uchebn. Zaved., Tsvetn. Metall., 2, 1976, 130-132.

[208] Rao, V.B., Houska, C.R., Metallurgical Transactions A, 14[1] 1983, 61-66. https://doi.org/10.1007/BF02643738

[209] Slusser, G.S., Ryan, J.G., Shore, S.E., Lavoie, M.A., Sullivan, T.D., Journal of Vacuum Science and Technology A, 7[3] 1989, 1568-1572. https://doi.org/10.1116/1.576094

[210] Barron, S.C., Kelly, S.T., Kirchhoff, J., Knepper, R., Fisher, K., Livi, K.J.T., Dufresne, E.M., Fezzaa, K., Barbee, T.W., Hufnagel, T.C., Weihs, T.P., Journal of Applied Physics, 114[22] 2013, 223517. https://doi.org/10.1063/1.4840915

[211] Sen, S., Lake, M., Schaaf, P., Applied Surface Science, 2018, Article in Press.

[212] Vohra, M., Weihs, T.P., Knio, O.M., Combustion and Flame, 162[1] 2015, 249-257. https://doi.org/10.1016/j.combustflame.2014.07.010

Materials Research Forum LLC
doi: http://dx.doi.org/10.21741/9781644900093

[213] Overdeep, K.R., Joress, H., Zhou, L., Livi, K.J.T., Barron, S.C., Grapes, M.D., Shanks, K.S., Dale, D.S., Tate, M.W., Philipp, H.T., Gruner, S.M., Hufnagel, T.C., Weihs, T.P., Combustion and Flame, 191, 2018, 442-452. https://doi.org/10.1016/j.combustflame.2017.11.023

[214] Overdeep, K.R., Schmauss, T.A., Panigrahi, A., Weihs, T.P., Combustion and Flame, 196, 2018, 88-98. https://doi.org/10.1016/j.combustflame.2018.05.035

[215] Overdeep, K.R., Livi, K.J.T., Allen, D.J., Glumac, N.G., Weihs, T.P., Combustion and Flame, 162[7] 2015, 2855-2864. https://doi.org/10.1016/j.combustflame.2015.03.023

[216] Sen, S., Lake, M., Schaaf, P., Surface and Coatings Technology, 340, 2018, 66-73. https://doi.org/10.1016/j.surfcoat.2018.02.014

[217] Fisher, K., Barron, S.C., Bonds, M.A., Knepper, R., Livi, K.J.T., Campbell, G.H., Browning, N.D., Weihs, T.P., Journal of Applied Physics, 114[24] 2013, 243509. https://doi.org/10.1063/1.4850915

[218] Laik, A., Bhanumurthy, K., Kale, G.B., Journal of Nuclear Materials, 305[2-3] 2002, 124-133. https://doi.org/10.1016/S0022-3115(02)01028-0

[219] Laik, A., Bhanumurthy, K., Kale, G.B., Intermetallics, 12[1] 2004, 69-74. https://doi.org/10.1016/j.intermet.2003.09.002

[220] Adams, D.P., Rodriguez, M.A., McDonald, J.P., Bai, M.M., Jones, E., Brewer, L., Moore, J.J., Journal of Applied Physics, 106[9] 2009, 093505. https://doi.org/10.1063/1.3253591

[221] Adams, D.P., Bai, M.M., Rodriguez, M.A., Moore, J.J., Brewer, L.N., Kelley, J.B., Proceedings of the 3rd International Brazing and Soldering Conference, 2006, 2006, 298-302.

[222] Suryanarayana, C., Moore, J.J., Radtke, R.P., Advanced Materials Processing, 159, 2001, 29.

[223] Wang, J., Besnoin, E., Duckham, A., Spey, S.J., Reiss, M.E., Knio, O.M., Weihs, T.P., Journal of Applied Physics, 95[1] 2004, 248-256. https://doi.org/10.1063/1.1629390

[224] Knepper, R., Fritz, G., Weihs, T.P., Journal of Materials Research, 23, 2008, 2009.

[225] Simões, S., Ramos, A.S., Viana, F., Vieira, M.T., Vieira, M.F., Metals, 6[5] 2016, 96. https://doi.org/10.3390/met6050096

[226] Simões, S., Viana, F., Ramos, A.S., Vieira, M.T., Vieira, M.F., Journal of Materials Science, 48[21] 2013, 7718-7727. https://doi.org/10.1007/s10853-013-7592-2

[227] Cavaleiro, A.J., Ramos, A.S., Martins, R.M.S., Fernandes, F.M.B., Morgiel, J., Baehtz, C., Vieira, M.T.F., Journal of Alloys and Compounds, 646, 2015, 1165-1171. https://doi.org/10.1016/j.jallcom.2015.06.037

[228] Tadayyon, S.M., Yoshinara, O., Tanaka, K., Japanese Journal of Applied Physics - 1, 31[7] 1992, 2226-2232.

[229] Mâaza, M., Sella, C., Ambroise, J.P., Kâabouchi, M., Milôche, M., Wehling, F., Groos, M., Journal of Applied Crystallography, 26[3] 1993, 334-342. https://doi.org/10.1107/S0021889892010355

[230] Efimenko, L.P., Petrova, L.P., Russian Metallurgy, [5] 1998, 145-150.

[231] Meng, W.J., Fultz, B., Ma, E., Johnson, W.L., Applied Physics Letters, 148, 1987, 148-156.

[232] Bokshtein, S.Z., Legk. Splavy Metody Ikh Obrab., 1968, 254-261.

[233] Nakajima, H., Maekawa, S., Aoki, Y., Koiwa, M., Transactions of the Japan Institute of Metals, 26[1] 1985, 1-6. https://doi.org/10.2320/matertrans1960.26.1

[234] Bouhki, M., Bruson, A., Guilmin, P., Solid State Communications, 79[5] 1991, 389-393. https://doi.org/10.1016/0038-1098(91)90490-M

[235] Barron, S.C., Knepper, R., Walker, N., Weihs, T.P., Journal of Applied Physics, 109[1] 2011, 013519 https://doi.org/10.1063/1.3527925

[236] McDonald, J.P., Rodriguez, M.A., Jones, E.D., Adam, D.P., Journal of Materials Research, 25[4] 2010, 718-727. https://doi.org/10.1557/JMR.2010.0091

[237] Reeves, R.V. Rodriguez, M.A. Jones, E.D., Adams, D.P., Journal of Physical Chemistry C, 116[33] 2012, 17904-17912. https://doi.org/10.1021/jp303785r

[238] Reeves, R.V., Rodriguez, M.A., Jones, E.D., Adams, D.P., Materials Research Society Symposium Proceedings, 1405, 2011, 90-95.

[239] Kim, K., Metals and Materials International, 23[2] 2017, 326-335. https://doi.org/10.1007/s12540-017-6379-4

[240] Zhang, Y., Jiang, H., Xing, D., Zhao, X., Zhang, W., Li, Y., Applied Thermal Engineering, 117, 2017, 617-621. https://doi.org/10.1016/j.applthermaleng.2016.10.070

[241] Koundinyan, S.P., Bdzil, J.B., Matalon, M., Stewart, D.S., Combustion and Flame, 162[12] 2015, 4486-4496. https://doi.org/10.1016/j.combustflame.2015.08.023

[242] Lee, D., Sim, G.D., Xiao, K., Vlassak, J.J., Journal of Physical Chemistry C, 118[36] 2014, 21192-21198. https://doi.org/10.1021/jp505941g

Materials Research Forum LLC
doi: http://dx.doi.org/10.21741/9781644900093

[243] Reiss, M.E., Esber, C.M., Van Heerden, D., Gavens, A.J., Williams, M.E., Weihs, T.P., Materials Science and Engineering A, 261[1-2] 1999, 217-222. https://doi.org/10.1016/S0921-5093(98)01069-7

[244] Clevenger, L.A., Thompson, C.V., Tu, K.N., Journal of Applied Physics, 67, 1990, 2894. https://doi.org/10.1063/1.345429

[245] Clevenger, L.A., Thompson, C.V., Cammarata, R.C., Tu, K.N., Applied Physics Letters, 52[10] 1998, 795-797. https://doi.org/10.1063/1.99644

[246] Kissenger, H.E., Analytical Chemistry, 29[11] 1957, 1702-1706 https://doi.org/10.1021/ac60131a045

[247] Braeuer, J., Besser, J., Wiemer, M., Gessner, T., 16th International Solid-State Sensors, Actuators and Microsystems Conference, 2011, 1332-1335.

[248] Sen, S., Lake, M., Schaaf, P., Vacuum, 156, 2018, 205-211. https://doi.org/10.1016/j.vacuum.2018.07.033

[249] Sen, S., Lake, M., Grieseler, R., Schaaf, P., Surface and Coatings Technology, 327, 2017, 25-31. https://doi.org/10.1016/j.surfcoat.2017.07.065

[250] Floro, J.A., Journal of Vacuum Science & Technology A, 4, 1986, 631-636. https://doi.org/10.1116/1.573848

[251] Piekiel, N.W., Morris, C.J., ACS Applied Materials and Interfaces, 7[18] 2015, 9889-9897. https://doi.org/10.1021/acsami.5b01964

Keyword Index

Materials Research Forum LLC
doi: http://dx.doi.org/10.21741/9781644900093